INFORMATION MECHANICS
Transformation of Information in Management, Command, Control and Communication

MATHEMATICS AND ITS APPLICATIONS

Series Editor: G. M. BELL, Professor of Mathematics,
King's College London (KQC), University of London

NUMERICAL ANALYSIS, STATISTICS AND OPERATIONAL RESEARCH
Editor: B. W. CONOLLY, Professor of Mathematics (Operational Research),
Queen Mary College, University of London

Mathematics and its applications are now awe-inspiring in their scope, variety and depth. Not only is there rapid growth in pure mathematics and its applications to the traditional fields of the physical sciences, engineering and statistics, but new fields of application are emerging in biology, ecology and social organization. The user of mathematics must assimilate subtle new techniques and also learn to handle the great power of the computer efficiently and economically.

The need for clear, concise and authoritative texts is thus greater than ever and our series will endeavour to supply this need. It aims to be comprehensive and yet flexible. Works surveying recent research will introduce new areas and up-to-date mathematical methods. Undergraduate texts on established topics will stimulate student interest by including applications relevant at the present day. The series will also include selected volumes of lecture notes which will enable certain important topics to be presented earlier than would otherwise be possible.

In all these ways it is hoped to render a valuable service to those who learn, teach, develop and use mathematics.

Mathematics and its Applications
Series Editor: G. M. BELL, Professor of Mathematics, King's College London (KQC), University of London

Anderson, I.	Combinatorial Designs
Artmann, B.	The Concept of Number
Arczewski, K. & Pietrucha, J.	Mathematical Modelling in Discrete Mechanical Systems
Arczewski, K. and Pietrucha, J.	Mathematical Modelling in Continuous Mechanical Systems
Bainov, D.D. & Konstantinov, M.	The Averaging Method and its Applications
Baker, A.C. & Porteous, H.L.	Linear Algebra and Differential Equations
Balcerzyk, S. & Joszefiak, T.	Commutative Rings
Balcerzyk, S. & Joszefiak, T.	Noetherian and Krull Rings
Baldock, G.R. & Bridgeman, T.	Mathematical Theory of Wave Motion
Ball, M.A.	Mathematics in the Social and Life Sciences: Theories, Models and Methods
de Barra, G.	Measure Theory and Integration
Bartak, J., Herrmann, L., Lovicar, V. & Vejvoda, D.	Partial Differential Equations of Evolution
Bell, G.M. and Lavis, D.A.	Co-operative Phenomena in Lattice Models, Vols. I & II
Berkshire, F.H.	Mountain and Lee Waves
Berry, J.S., Burghes, D.N., Huntley, I.D., James, D.J.G. & Moscardini, A.O.	Mathematical Modelling Courses
Berry, J.S., Burghes, D.N., Huntley, I.D., James, D.J.G. & Moscardini, A.O.	Mathematical Methodology, Models and Micros
Berry, J.S., Burghes, D.N., Huntley, I.D., James, D.J.G. & Moscardini, A.O.	Teaching and Applying Mathematical Modelling
Blum, W.	Applications and Modelling in Learning and Teaching Mathematics
Brown, R.	Topology
Burghes, D.N. & Borrie, M.	Modelling with Differential Equations
Burghes, D.N. & Downs, A.M.	Modern Introduction to Classical Mechanics and Control
Burghes, D.N. & Graham, A.	Introduction to Control Theory, including Optimal Control
Burghes, D.N., Huntley, I. & McDonald, J.	Applying Mathematics
Burghes, D.N. & Wood, A.D.	Mathematical Models in the Social, Management and Life Sciences
Butkovskiy, A.G.	Green's Functions and Transfer Functions Handbook
Cartwright, M.	Fourier Methods: Applications in Mathematics, Engineering and Science
Cerny, I.	Complex Domain Analysis
Chorlton, F.	Textbook of Dynamics, 2nd Edition
Chorlton, F.	Vector and Tensor Methods
Cohen, D.E.	Computability and Logic
Crapper, G.D.	Introduction to Water Waves
Cross, M. & Moscardini, A.O.	Learning the Art of Mathematical Modelling
Cullen, M.R.	Linear Models in Biology
Dunning-Davies, J.	Mathematical Methods for Mathematicians, Physical Scientists and Engineers
Eason, G., Coles, C.W. & Gettinby, G.	Mathematics and Statistics for the Bio-sciences
El Jai, A. & Pritchard, A.J.	Sensors and Controls in the Analysis of Distributed Systems
Exton, H.	Multiple Hypergeometric Functions and Applications

Series continued at back of book

UNIVERSITY OF MAINE

RAYMOND H. FOGLER LIBRARY

INFORMATION MECHANICS
Transformation of Information in Management, Command, Control and Communication

BRIAN CONOLLY, B.A., M.A.
Professor of Mathematics (Operational Research)
Queen Mary College, University of London

and

JOHN G. PIERCE, B.S., M.S., Ph.D.
Senior Research Scientist
Radix Systems Inc., Rockville, Maryland, USA

ELLIS HORWOOD LIMITED
Publishers · Chichester

Halsted Press: a division of
JOHN WILEY & SONS
New York · Chichester · Brisbane · Toronto

First published in 1988 by
ELLIS HORWOOD LIMITED
Market Cross House, Cooper Street,
Chichester, West Sussex, PO19 1EB, England
The publisher's colophon is reproduced from James Gillison's drawing of the ancient Market Cross, Chichester.

Distributors:
Australia and New Zealand:
JACARANDA WILEY LIMITED
GPO Box 859, Brisbane, Queensland 4001, Australia

Canada:
JOHN WILEY & SONS CANADA LIMITED
22 Worcester Road, Rexdale, Ontario, Canada

Europe and Africa:
JOHN WILEY & SONS LIMITED
Baffins Lane, Chichester, West Sussex, England

North and South America and the rest of the world:
Halsted Press: a division of
JOHN WILEY & SONS
605 Third Avenue, New York, NY 10158, USA

South-East Asia
JOHN WILEY & SONS (SEA) PTE LIMITED
37 Jalan Pemimpin # 05–04
Block B, Union Industrial Building, Singapore 2057

Indian Subcontinent
WILEY EASTERN LIMITED
4835/24 Ansari Road
Daryaganj, New Delhi 110002, India

© 1988 B. Conolly and J. H. Pierce/Ellis Horwood Limited

British Library Cataloguing in Publication Data
Conolly, Brian
Information mechanic: transformation of information in management,
command, control and communication. —
(Ellis Horwood series in mathematics and its applications).
1. Information theory
I. Title II. Pierce, John H., *1936–*
001.53'9

Library of Congress Card No. 88-9377

ISBN 0–7458–0392–X (Ellis Horwood Limited)
ISBN 0–470–21136–9 (Halsted Press)

Typeset in Times by Ellis Horwood Limited
Printed in Great Britain by Unwin Bros., Woking

COPYRIGHT NOTICE
All Rights Reserved. No part of this publication may be reproduced, stored in a retrieval system, or transmitted, in any form or by any means, electronic, mechanical, photocopying, recording or otherwise, without the permission of Ellis Horwood Limited, Market Cross House, Cooper Street, Chichester, West Sussex, England.

Table of contents

Preface 7

1 Simple models for search in the presence of false targets
1.1 Concept of classification. 13
1.2 Two cell search: simple false target structure 18
1.3 Two cell search: single target and arbitrary numbers of false targets. . . .31
1.4 Two cell search: two targets and arbitrary numbers of false targets 37

2 Information provided by regular surveillance of a moving target 48

3 Merger of data in a filter centre
3.1 Basic concepts 56
3.2 Generalized theory 69

4 Management of an information channel with a priority facility
4.1 Introduction to 'head-of-the-line' priority. 87
4.2 The mechanism 88
4.3 Statistical description of system state. .88
4.4 Non-priority system time: general discussion 90
4.5 General form of probability densities of system time 92
4.6 Processing of equations (166–171) and the explicit form of $g_p(t)$ 94
4.7 Some statistics of system time and system state 97
4.8 Non-priority system state and the form of $g_{np}(t)$ 100
4.9 Management of a priority system 112

5 Single-channel service with alternating priority
5.1 Introduction to alternating priority 118
5.2 Statistical description of system state. 118
5.3 A generating function approach to the state probabilities 120
5.4 System state performance 126
5.5 System time including processing 130

6 Modified Lanchester equations incorporating effects of information
    6.1 Introduction 135
    6.2 Analysis 138
    6.3 Numerical studies 145
    6.4 Summary 162

Appendices
Appendix A: Technical notes on text of chapters 4 and 5 165
Appendix B: Calculation of $A(x,y)$, $B(x,y)$ for $x, y \; \varepsilon (0,1)$ 168
Appendix C: Some values of the standard deviations for alternating priority .. 170

References 171

Index 173

Preface

The studies in this book are intended to illuminate some of the processes of concern in military command, control, and communication (often referred to by the acronymic C^3):search; occasional observation and surveillance; transmission of information over possibly faulty channels; aspects of priority in message traffic; interaction between competitors who gain access to new information. C^3 is often called the last unsolved problem of military operational research. It is, of course, a tangle of interrelated problems, not just one, and its equivalent in business and industry presents no fewer enigmas and challenges. The nomenclature may be different, but to each problem in the military context there exists at least one non-military counterpart.

The problems of command and control arise in the context of competition between organizations. A hierarchy of decision-makers on each side collects and processes information concerning the deployment of own resources and their utilization, and seeks, scrutinizes, and interprets information about the opponents as a basis for gaining tactical and strategic advantage. Information may be sought by agents and communicated by observers: the messages may be delayed, misconstrued, and corrupted in transmission. The evolution of the decision-maker's strategy, involving processes which are incompletely understood and the subject of much investigation by psychologists and their acolytes, is a continually varying process influenced by, and influencing, the interactions between his own organization and the competitors.

The formal study of military command and control as a scientific discipline has developed only in the past decade [20]. It is still young, and the appropriate theoretical formulations continue to be the subject of considerable debate. A key issue in that debate is the choice among existing mathematical formalisms as the proper vehicle for the description of the phenomena, or, indeed, whether existing formalisms are adequate. Among the formalisms proposed are control theory, information theory, fuzzy set theory, and catastrophe theory. The first two of these have appeal at a fundamental theoretical level: the latter two at a phenomenological level.

Control theory has a natural appeal because of the structural similarity of military command and control to the problems normally addressed by control engineering. A process (the battle) has some objective; deviations from the objective are measured; corrective action is then signalled through a feedback loop. In practice, the situations encountered in military command and control are extremely complex, and the ability of the commander to effect corrective action is very uncertain. Nonetheless, the structural analogy is valid, and some limited success has been achieved by applying the concepts and techniques of control theory to military command and control.

Information theory also has a natural appeal, but for quite different reasons. There is no structural analogy; rather, information, in the Shannon sense, is the fundamental quantity that pervades the entire process of command and control. Information is gathered by battlefield sensors; information is corrupted by noisy communication links; information decays as time changes the reality of the battlefield; information is the basis of command decisions; and information, again corrupted by noise, is the essence of directives from the commander to his combat forces.

Fuzzy set theory has been employed to express the inherent imprecision that is characteristic of most command and control problems, but its basis is controversial and it is far from universally accepted (see [1] for example). Catastrophe theory has been introduced, largely on an *ad hoc* basis, because certain sudden changes in the course of a battle that are related to the breakdown of the command and control apparatus appear to resemble some of the catastrophes defined by Thom [2].

The analytical studies collected in this book do not attempt to resolve the issue of which is the correct mathematical framework for the study of command and control. Experience, however, tells the seasoned analyst to keep an open mind. This being the case we must say that we have found information theory to be a promising and widely applicable linking agent.

Attempts to apply information theory to various military operations embedded in the command and control process are hardly new, but earlier efforts met with indifferent success. Regarding the process of search for a target, Koopman [3] stated: '. . . ever since the mid-nineteen-forties when the theories of information and search became subjects of general interest, attempts have been made to apply the theory of information to problems of search. These have proved disappointing: neither the formulas (sic) nor the concepts of information theory have found a place in clarifying the problems of search.'

A major contributing factor to this pessimistic view was the 1961 paper by Mela [4]. This gives numerical examples for simple search models, demonstrating that a search policy designed to maximize detection probability does not necessarily maximize either the gain in expected information or the probability of correctly committing forces on the basis of search outcomes. He concluded: '. . . it does not seem likely that there is any intimate connection between search theory and information theory.'

Despite this apparently negative assertion, other researchers continued to probe information-theoretic approaches, often obtaining results that contradicted those of Mela.

In 1973, Richardson [5] conducted a simulation study to explore alternative surveillance policies in a false- and moving-target environment. One of his policy options was a maximum information gain policy. He concluded:

> '. . . the maximum information gain policy appears to have desirable characteristics in the idealised surveillance scenario considered. Among these characteristics are . . . good initial behavior in the early stages (of surveillance) and good asymptotic behavior in the later stages.'

As a result of Richardson's findings, Barker [6] was able to prove a theorem that stated that 'subject to a constraint on total search effort, the allocation of search effort that maximises the probability of detection also maximises the entropy of the posterior distribution.' This theorem appeared to contradict the numerical results of Mela.

This contradictory situation was clarified in 1978 by Pierce [7]. In that paper, Pierce showed, among other things, that: (1) there are many different information or entropy quantities that can be calculated in a search situation, and that Barker's theorem referred to a different quantity from that calculated by Mela; and (2) Mela's examples included the single exception to an otherwise general rule that relates information gain and search effectiveness. The results showed that the connection between information theory and optimal search is strong, but not universal. Two basic conclusions are:

> 'The search policy that maximizes the probability of detection is the one that uses up the information contained in the prior ensemble at the maximum rate.'

> 'In a wide range of practical cases (but not all cases) the search policy that maximizes the detection probability is the one that creates expected information at the maximum rate. We achieve the maximum rate of creation by using the existing information of the prior ensemble as fast as possible, thereby gaining an early detection, with its concomitantly large increase of expected information.'

The work described in this book includes some extensions to this earlier work on information theory and search. It also covers applications of information-theoretic ideas to other processes that are part of the whole command and control process.

Chapter 1, an extension of [7], discusses the relationship of information theory and search when false targets are present. For those unfamiliar with the nomenclature, a false target is a contact that passes most of the tests that a true target would be required to satisfy, and yet is not a true target. Think, for example, of a whale and a submarine. Search theory has widely recognised applications in a non-military context and, indeed, was used early on in connection with prospecting for mineral deposits, in particular, oil. See references in Chapter 3 of Conolly [8], and the source book, Koopman [9], by the father of the subject. False contacts, or leads, bedevil such search, and since search implies the use of resources that might be deployed to greater advantage elsewhere it is important to minimize the wastage and to arrive quickly at reliable conclusions. Search for new markets for a range of products is another example where the gathering of possibly misleading information and the misrepresentation of correct information may be critical.

Chapter 2 develops a mathematical description of the information generated by periodic surveillance of a moving target, in which detection of the target is not an issue. The loss of information due to movement of the target, and the gain acquired from measurement, are quantified. The methodology is ripe for development and its application to, for instance, certain types of police operation, anti-terrorist activity and the interdiction of smuggling, as well as to military objectives, is evident.

Chapter 3 treats the merger, in a filter centre, of data about the same situation from various sources. The effects of erroneous or missing data are described in information-theoretic terms. The models developed, and their extensions, are of interest to communication engineers. They are concerned with the transmission of information from an observer to an operational centre where, in real life, information from different sources concerning a complex situation is collected and reconciled. In spite of the best efforts of the communication engineers, contradictory information may be received, and this creates confusion and impedes effective decision-making. Different methods of information transmission may be used to mitigate the distortion of information passed between transmission and reception, and this is the theme of the chapter, attempting to understand the trade-offs between different methods of information transmission, modelled here as noisy and fallible channels. An underlying issue is how best to interpret faulty data.

Chapters 4 and 5 examine, in terms of the formalism of queueing theory, the effects of assigning priority to one class of information to be transmitted over a link with finite capacity. This is a summary of simple but not particularly well-known theory which has been used previously in this application: but, again, in the context of other than message traffic it is ready to hand. Emergency centres in hospitals provide an apt example. The thrust of this discussion is the damage that can follow from careless application of priority. A survey is given of the crucial operational factors which must be controlled.

Chapter 6 has its roots in models created by Lanchester [10]† to describe the evolution in deterministic terms of what he called ancient and modern warfare. No account was taken in these early models of the influence that might bear on the fortunes of the antagonists if effort was spent on gathering information and exploiting it. This is the first known attempt to model such a situation, and it does exhibit features known to arise in experience. The setting need not be military. Lanchester-type modelling belongs to the family of birth and death differential equation models applied to competition in a wide variety of contexts, from ecology to advertising.

We repeat, in conclusion, that the concept of information plays a central role in these studies. The related measure, entropy, is invoked specifically, and it is recalled that the principle of maximizing entropy has had considerable success in statistical estimation procedures of which spectral estimation provides a striking example (Burg [11]). A recent symposium which exploited related ideas in the context of models for information technology is reported in [25].

The studies in this book were carried out as part of a research programme at the

---

† The relevant chapters are reprinted with commentary in [17]. See also [8], Chapter 1, for a discussion.

Saclant ASW Research Centre. Grateful acknowledgement is made to the Director, Dr R. R. Goodman, for permission to make them available to a wider audience.

Brian Conolly and John G. Pierce
December 1987

1
Simple models for search in presence of false targets

1.1 CONCEPT OF CLASSIFICATION

1.1.1 The model

The search model developed in this study introduces the concept of classification, that is the capability of identifying an object, rightly or wrongly. If a genuine target T is detected we shall say that with probability c it is declared correctly to be T, and with probability $\bar{c}(=1-c)$ it is declared to be $\bar{\bar{T}}$ (not T), that is to say F (false). Likewise, if a false target is detected there is a probability c that it is correctly identified as F, and $\bar{c}$ that it is stated to be T. The bar is commonly used to denote the logical complement of an event, or 1 minus the probability of the event.

We shall consider search in a single area, or cell.

Underlying the whole endeavour is the attempt to measure, accepting the limitations and idealization of the model, the corrupting effect of false targets on the information made available by the search of an area.

In the classical formulation of search theory a sensor-carrying vehicle can be thought of as stationary at the centre of a search area. The sensor provides some kind of display of the area and this will, in general, be littered with multiple target-like images. Note that the stationary sensor concept corresponds to surveillance, but that the same machinery can be made to apply to search by a moving vehicle in certain cases by placing it at relative rest.

We shall limit this introductory analysis to the case where, during the time of the search, the area may contain one genuine target T with probability t, and one target-like false target F with probability f. The two may coexist and are taken to be independent. Four fundamental subsets in the event space are thus $TF, T\bar{F}, \bar{T}F, \bar{T}\bar{F}$, with associated probabilities $tf, t\bar{f}, \bar{t}f, \bar{t}\bar{f}$, where $t+\bar{t}=1, f+\bar{f}=1$ $(0 \leq t, f \leq 1)$. The probability that T/F is detected is denoted by d_t/d_f. There are 16 cases to distinguish. These are listed in Table 1.

It is useful to set the cases out in detail and to check that the sum of the

Table 1 — Event space and probabilities

Set	Case	Probability	Search outcome	Joint event
I(TF)	1	$tf\, d_t\, d_f\, c^2$	Y_1	(X_1, Y_1)
	2	$tf\, d_t\, d_f\, c\bar{c}$	Y_1	(X_1, Y_1)
	3	$tf\, d_t\, d_f\, \bar{c}c$	Y_2	(X_1, Y_2)
	4	$tf\, d_t\, d_f\, \bar{c}^2$	Y_1	(X_1, Y_1)
	5	$tf\, d_t\, \bar{d}_f\, c$	Y_1	(X_1, Y_1)
	6	$tf\, d_t\, \bar{d}_f\, \bar{c}$	Y_2	(X_1, Y_2)
	7	$tf\, \bar{d}_t\, d_f\, c$	Y_2	(X_1, Y_2)
	8	$tf\, \bar{d}_t\, d_f\, \bar{c}$	Y_1	(X_1, Y_1)
	9	$tf\, \bar{d}_t\, \bar{d}_f$	Y_0	(X_1, Y_0)
II(T$\bar{\text{F}}$)	10	$t\bar{f}\, d_t\, c$	Y_1	(X_1, Y_1)
	11	$t\bar{f}\, d_t\, \bar{c}$	Y_2	(X_1, Y_2)
	12	$t\bar{f}\, \bar{d}_t$	Y_0	(X_1, Y_0)
III($\bar{\text{T}}$F)	13	$\bar{t}f\, d_f\, c$	Y_2	(X_0, Y_2)
	14	$\bar{t}f\, d_f\, \bar{c}$	Y_1	(X_0, Y_2)
	15	$\bar{t}f\, \bar{d}_f$	Y_0	(X_0, Y_0)
IV($\bar{\text{T}}\bar{\text{F}}$)	16	$\bar{t}\bar{f}$	Y_0	(X_0, Y_0)

probabilities is unity. The notation is intended to be self-explanatory. For instance, case 1 is the event that both T and F are present, both detected, and both correctly identified. Case 2 is the event that both T and F are present and both detected, while T is correctly identified and F is not: in this case F would be thought to be a T, but the conclusion that T is present is true. If we took into account the loss in useful effort consequent on the prosecution of a false target this would be detrimental. Again, case 4 would lead to the correct conclusion that T is present, but for the wrong reason. Case 14 belongs to the same category.

Possible states of nature are:

X_0 : no T present
X_1 : T present

Possible outcomes of the search are:

Y_0 : no detection
Y_1 : at least one detection made and the conclusion drawn that X_1 is the state of nature
Y_2 : at least one detection made and the conclusion drawn that X_0 is the state of nature

The cases giving rise to Y_0, Y_1, Y_2 are then:

Sec. 1.1]	**Concept of classification**	15

Y_0 : 9, 12, 15, 16
Y_1 : 1, 2, 4, 5, 8, 10, 14
Y_2 : 3, 6, 7, 11, 13

From the information-theoretic viewpoint there are six fundamental joint probabilities to assemble: these are listed below with the corresponding contributions from Table 1.

$$P(X_0, Y_0) = 15 + 16$$
$$P(X_1, Y_0) = 9 + 12$$
$$P(X_0, Y_1) = 14$$
$$P(X_1, Y_1) = 1 + 2 + 4 + 5 + 8 + 10$$
$$P(X_0, Y_2) = 13$$
$$P(X_1, Y_2) = 3 + 6 + 7 + 11$$

For the record, these give:

$$P(X_0, Y_0) = \bar{t}(1 - f\,d_f) \tag{1}$$
$$P(X_1, Y_0) = t\,\bar{d_t}(1 - f\,d_f) \tag{2}$$
$$P(Y_0) = (1 - t\,d_t)(1 - f\,d_f) \tag{3}$$
$$P(X_0|Y_0) = \bar{t}/(1 - t\,d_t) \tag{4}$$
$$P(X_1|Y_0) = t\,\bar{d_t}/(1 - t\,d_t) \tag{5}$$
$$P(X_0, Y_1) = \bar{t}\,f\,d_f\,\bar{c} \tag{6}$$
$$P(X_1, Y_1) = t(d_t\,c + f\,d_f\,\bar{c}\,(1 - c\,d_t)) \tag{7}$$
$$P(Y_1) = t\,d_t\,c + f\,d_f\,\bar{c}\,(1 - t\,d_t\,c) \tag{8}$$
$$P(X_0|Y_1) = P(X_0, Y_1)/P(Y_1), \quad P(X_1|Y_1) = P(X_1, Y_1)/P(Y_1) \tag{9}$$
$$P(X_0, Y_2) = \bar{t}\,f\,d_f\,c \tag{10}$$
$$P(X_1, Y_2) = t(\bar{f}\,d_t\,\bar{c} + f(d_f\,c(1 - d_t\,c) + \bar{d_f}\,d_t\,\bar{c})) \tag{11}$$
$$P(Y_2) = P(X_0, Y_2) + P(X_1, Y_2) \tag{12}$$
$$P(X_0|Y_2) = P(X_0, Y_2)/P(Y_2),$$
$$P(X_1|Y_2) = P(X_1, Y_2)/P(Y_2) . \tag{13}$$

The explicit expressions are readily capable of interpretation and could have been written down immediately from first principles.

Finally, the increase in information about X_j resulting from search outcome Y_k, is

$$\log P(X_j|Y_k) - \log P(X_j) \tag{14}$$

and the mean value I_E over all possibilities is

$$I_E = \sum_{j=0}^{1} \sum_{k=0}^{2} P(X_j, Y_k) \log(P(X_j|Y_k)/P(X_j)) . \tag{15}$$

This will be tabulated in the next paragraph for a range of parameter values.

1.1.2 Behaviour of I_E

(a) Degradation of I_E as f increases

Intuitively one expects I_E to decrease as f increases, the values of the remaining parameters being fixed. Table 2 is an example.

Table 2 — I_E as a function of f
$d_t = d_f = 0.5$
$c = 0.5, t = 0.5$

f	I_E
0.1	0.1547
0.3	0.1027
0.5	0.0738
0.7	0.0548
0.9	0.0413

(b) Variation of I_E with respect to t and f

For fixed f it appears that I_E has a maximum near $t = 0.5$. This would mean that the search provides on average the greatest information gain when, a priori, no preferential statement can be made about whether or not T is present. Less information is yielded when the prior belief in the presence or absence of T is strong. This is illustrated in Table 3.

Table 3 — I_E as a function of f and t
$d_t = d_f = 0.5, c = 0.5$

f	0.1	0.5	0.9
t	I_E	I_E	I_E
0.1	0.0717	0.0279	0.0147
0.2	0.1167	0.0488	0.0262
0.3	0.1433	0.0632	0.0344
0.4	0.1552	0.0715	0.0395
0.5	0.1547	0.0738	0.0413
0.6	0.1431	0.0704	0.0398
0.7	0.1214	0.0613	0.0351
0.8	0.0901	0.0466	0.0270
0.9	0.0496	0.0262	0.0153

(c) Variation of I_E with respect to c and t

It appears that I_E decreases to a minimum as c increases, and that it then increases again. This can be interpreted as meaning that if one has high certainty about one's power of identification, or great uncertainty, the information gain is greater than otherwise. This statement needs to be viewed against the probability of drawing the correct conclusion from a search. This is, of course, c. Table 4 provides an

Table 4 — I_E as a function of t and c
$d_t = d_f = 0.5, f = 0.5$

t/c	0.1	0.5	0.9
0.1	0.0573	0.0279	0.0771
0.3	0.1122	0.0632	0.1467
0.5	0.1202	0.0738	0.1545
0.7	0.0939	0.0613	0.1193
0.9	0.0383	0.0262	0.0481

illustration.

(d) Variation of I_E with d_t ($= d_f$) and c

Increase in d_t (here $= d_f$) procures a steady increase in I_E for each c. This is intuitively correct. Table 5 provides an illustration. It is to be noted that increasing d_t and d_f may be interpreted as equivalent to increasing the search effort.

Table 5 — I_E as a function of c and d_t with $d_t = d_f$
$f = 0.5, \quad t = 0.5$

c	0.1	0.5	0.9
$d_t = d_f$	I_E	I_E	I_E
0.1	0.0248	0.0133	0.0259
0.5	0.1202	0.0738	0.1545
0.9	0.2255	0.1660	0.3668

1.1.3 Behaviour of posterior probabilities

In addition to examining I_E we should also investigate posterior probabilities. Several of these are candidates, but one of great importance is $P(X_1|Y_1 + Y_2)$, the probability that the state of nature is X_1 (a target is present) given that the search has yielded a detection, and whatever the conclusion that was drawn therefrom. We have

$$P(Y_1 + Y_2) = t\,d_t + f\,d_f\,(1 - t\,d_t),$$

which is increased by the presence of a false target and is independent of prowess in identification. Then

$$P(X_1|Y_1 + Y_2) = \frac{t\,(1 - \bar{d}_t\,(1 - f\,d_f))}{t\,d_t + f\,d_f\,(1 - t\,d_t)}.$$

Before investigating this further we shall consider in Section 1.2 the partition of effort between two cells with the model structure of Section 1.1.

1.2 TWO CELL SEARCH: SIMPLE FALSE TARGET STRUCTURE

1.2.1 The model

Section 1.1 is extended to analyse the search in two cells. This is somewhat more complex and we shall, accordingly, provide a detailed account of the assumptions and arguments. The cells are designated C1 and C2, and it is supposed that C1 is the cell to be searched first.

(a) A target T is known to be either in C1 or C2. The probability that it lies in C1 is t; consequently the probability that it is in C2 is $\bar{t} = 1 - t$.

(b) There may be *one* false target (F) in either C1, C2, both, or neither; that is, we are assuming possible false target patterns FF, $\overline{F}F$, $F\overline{F}$, $\overline{FF}$ with respective probabilities f^2, $\bar{f}f$, $f\bar{f}$, $\bar{f}^2$. f is the probability of a false target. Extension to an arbitrary number of false targets is the subject of Section 1.3.

(c) Detection and classification are dealt with by an extension of the classical theory. Let us suppose that the cell searched contains both T and F. The searching vehicle deploys a sensor which is supposed to make repeated sweeps of the area. Each time the 'beam' sweeps over a target-like object constitutes a *detection opportunity* and, in the parlance of probability theory, is a Bernoulli trial whose outcome, by definition, is either success (detection), or failure (failure to detect). In this treatment the outcome of a trial is deemed to be independent of the outcome of previous trials. Let p_t, p_f be the probabilities of success when the object is, respectively, T and F, and let $q_t = 1 - p_t, q_f = 1 - p_f$. These should, in general, be variable from trial-to-trial, but we shall treat them as fixed, equivalent to replacing a sequence of variable probabilities by a mean value.

An object which causes the sensor to register 'detection' is then subjected to a classification or identification procedure which is assumed to absorb no search effort or time. The effect of a contrary assumption is investigated elsewhere. With probability c the object is correctly identified, and with probability $\bar{c} = 1 - c$ is stated to be what it is not. Thus an F which has been detected is with probability c identified as F, and with probability $\bar{c}$ is stated to be T.

1.2.2 Evaluation of probabilities

Let us suppose that the search effort applied permits N detection opportunities when a cell contains both T and F. The possible search outcomes are as follows:

(1) T detected first, correctly classified, search stops. The search outcome is Y_1: 'T seems to have been detected'.
(2) T detected first, incorrectly classified, search continues: F detected next, correctly classified, search proceeds to C2, denoted by the symbol '→2'.
(3) T detected first, incorrectly classified, search continues: F detected next, incorrectly classified and thought to be T. This leads to the correct conclusion that T is in the cell, but the wrong contact would be prosecuted, a fact that we do not take into account, search stops (Y_1).
(4) T detected first, incorrectly classified, search continues: F not detected (→2).
(5) F detected first, correctly classified, search continues: T detected next, correctly classified, search stops (Y_1).
(6) F detected first, correctly classified, search continues: T detected next, incorrectly classified (→2).
(7) F detected first, correctly classified, search continues, T not detected (→2).
(8) F detected first, incorrectly classified, search stops. This leads to the correct conclusion, but for the wrong reason, as in (3), search stops (Y_1).
(9) Neither T nor F detected (→2).
(10) Both T and F detected simultaneously. We shall see that this event can reasonably be neglected.

We now list the probabilities of (1) to (10). The assumption is that the search effort provides N detection opportunities.

Probability 1
Let T be detected at the nth trial ($1 \leq n \leq N$). Then

$$P(1) = c \sum_{n=1}^{N} q_f^n q_t^{n-1} p_t = c p_t q_f \frac{(1-(q_f q_t)^N)}{1-q_f q_t} \qquad (16)$$

Probability 2
Let T be detected at the nth trial ($1 \leq n \leq N$), and F at the $(n+r)$th trial $1 \leq r \leq N-n$). Then

$$P(2) = \sum_{n=1}^{N} q_f^n q_t^{n-1} p_t \bar{c} \sum_{r=1}^{N-n} q_f^{r-1} p_f c$$

$$= q_f p_t c \bar{c} \sum_{n=1}^{N} (q_f q_t)^{n-1} (1-q_f^{N-n})$$

$$= q_f p_t c\bar{c} \left[\frac{1-(q_f q_t)^N}{1-q_f q_t} - q_f^{N-1} \cdot \frac{1-q_t^N}{1-q_t} \right] \qquad (17)$$

Probability 3

$$P(3) = q_f p_t \bar{c}^2 \left[\frac{1-(q_f q_t)^N}{1-q_f q_t} - q_f^{N-1} \cdot \frac{1-q_t^N}{1-q_t} \right] \qquad (18)$$

Probability 4

$$P(4) = \sum_{n=1}^{N} q_f^n q_t^{n-1} p_t \bar{c}\, q_f^{N-n} = q_f^N\, \bar{c}\, (1-q_t^N) \qquad (19)$$

Probability 5

$$P(5) = \sum_{n=1}^{N} q_t^n q_f^{n-1} p_f c \sum_{r=1}^{N-n} q^{r-1t} p_t\, c$$

$$= q_t p_f c^2 \left[\frac{1-(q_f q_t)^N}{1-q_t q_f} - q_t^{N-1} \frac{1-q_f^N}{1-q_f} \right] \qquad (20)$$

Probability 6

$$P(6) = q_t p_f c\bar{c} \left[\frac{1-(q_f q_t)^N}{1-q_t q_f} - q_t^{N-1} \frac{1-q_f^N}{1-q_f} \right] \qquad (21)$$

Probability 7

$$P(7) = \sum_{n=1}^{N} q_t^n\, q_f^{n-1} p_f c q_t^{N-n} = q_t^N c(1-q_f^N) \qquad (22)$$

Probability 8

$$P(8) = \bar{c} \sum_{n=1}^{N} q_t^n q_f^{n-1} p_f = q_t \bar{c} p_f \frac{1-(q_t q_f)^N}{1-q_t q_f} \qquad (23)$$

Probability 9

$$P(9) = (q_t q_f)^N \qquad (24)$$

Probability 10

$$P(10) = \sum_{n=1}^{N} (q_t q_f)^{n-1} p_t p_f = p_t p_f \frac{(1-(q_t q_f)^N)}{(1-q_t q_f)} \qquad (25)$$

It is usual to suppose that the sensor scans fast, at time intervals h, say. If total finite time (or effort) z is allocated to the search, we have $z = Nh$. Since N Bernoulli trials with success probability p yield an average of Np successes, it is necessary to suppose p to be of the form $\alpha h + o(h)$ in order to have a finite mean as $N \to \infty$. This means that the probability of N failures, q^N, is given by $(1-\alpha h)^N = (1-\alpha h)^{z/h} \to \exp(-\alpha z)$ as $h \to 0$. On this basis we can now rewrite the probabilities $P(1)$ to $P(10)$ in exponential form. First we put

$$p_t = \alpha h + o(h), \qquad p_f = \beta h + o(h). \qquad (26)$$

Then, for example,

$$P(1) = c\alpha h(1-\beta h)\left[\frac{1-e^{-(\alpha+\beta)z}}{1-(1-\alpha h)(1-\beta h)}\right] + o(h) \to \frac{c\alpha(1-e^{-(\alpha+\beta)z})}{(\alpha+\beta)} \qquad (27)$$

as $h \to 0$.

This and other approximations to $P(n)$ can be obtained directly by supposing that T and F are members of two separate and independent Poisson streams with mean rates α and β respectively. Then

$$P(1) = \alpha c \int_0^z e^{-\alpha s} e^{-\beta s} ds = \frac{\alpha c}{(\alpha + \beta)} (1 - e^{-(\alpha + \beta)z}) \ . \tag{28}$$

Similarly we get

$$P(2) = \alpha c \bar{c} \left[\frac{1 - e^{-(\alpha + \beta)z}}{\alpha + \beta} - \frac{e^{-\beta z}(1 - e^{-\alpha z})}{\alpha} \right] \tag{29}$$

$$P(3) = \alpha \bar{c}^2 \left[\frac{1 - e^{-(\alpha + \beta)z}}{\alpha + \beta} - \frac{e^{-\beta z}(1 - e^{-\alpha z})}{\alpha} \right] \tag{30}$$

$$P(4) = \bar{c} \, e^{-\beta z} (1 - e^{-\alpha z}) \tag{31}$$

$$P(5) = \beta c^2 \left[\frac{1 - e^{-(\alpha + \beta)z}}{\alpha + \beta} - \frac{e^{-\alpha z}(1 - e^{-\beta z})}{\beta} \right] \tag{32}$$

$$P(6) = \beta c \bar{c} \left[\frac{1 - e^{-(\alpha + \beta)z}}{\alpha + \beta} - \frac{e^{-\alpha z}(1 - e^{-\beta z})}{\beta} \right] \tag{33}$$

$$P(7) = c \, e^{-\alpha z} (1 - e^{-\beta z}) \tag{34}$$

$$P(8) = \beta \bar{c} \left[\frac{1 - e^{-(\alpha + \beta)z}}{\alpha + \beta} \right] \tag{35}$$

$$P(9) = e^{-(\alpha + \beta)z} \ . \tag{36}$$

Finally, $P(10) = o(h)$ and is accordingly negligible in this approximation.

In the foregoing we have considered in detail the case where the first cell searched contains T and F. If it contains nothing the action must be to continue the search in another cell, or agree to an inconclusive outcome of the search.

The remaining cases are:

Cell content	Outcome	Probability	Classification	Action
T	Detected	$1 - e^{-\alpha z}$	Correct (c)	Stop
T	Detected	$1 - e^{-\alpha z}$	Incorrect $(\bar{c})$	Continue
T	Not detected	$e^{-\alpha z}$	——	Continue
F	Detected	$1 - e^{-\beta z}$	Correct (c)	Continue
F	Detected	$1 - e^{-\beta z}$	Incorrect $(\bar{c})$	Stop
F	Not detected	$e^{-\beta z}$	——	Continue

Sec. 1.2]   **Two sell search: simple false target structure**   23

The final assumption of this analysis is a search protocol. We have to designate a cell to be searched first and that will be defined to be cell 1 (C1). This is implicit in the foregoing analysis. To C1 will be allocated search effort z: to the other, $\bar{z} = 1 - z$. One of our problems is the optimal allocation of z. The search procedure is as follows: C1 is searched and if a target is detected, *or thought to be detected*, the search stops, rightly or wrongly, with the conclusion that the state of nature is X_1: = the target is in C1, whether or not there is a false target as well. To complement this is the state X_2 of nature making the corresponding statement about C2.

If the search in C1 yields no detection at all, or the detection, apparent or otherwise, of a false target, the search continues in C2. The searcher expects to find the target in C2 and perhaps a false tareget as well.

At the end of the search the following situations may obtain:

Y_0 : the target is not known to have been detected. One or more false targets may have been detected and correctly identified, or the target may itself have been detected, but misclassified.

Y_1 : an object has been declared to be the genuine target and the cell in which it has been identified is the correct one, even if for the wrong reason (misidentification of a false target).

Y_2 : as above, but the supposed target is declared to be in the cell where it is not.

The possible configurations awaiting the searcher are shown in the following table:

Cell 1	Probability	Cell 2	Probability
TF	tf	$F\bar{T}$	$\underline{f}$
TF	$t\underline{f}$	$\overline{FT}$	$\underline{f}$
T$\bar{F}$	$t\underline{f}$	$F\bar{T}$	$\underline{f}$
T$\bar{F}$	$t\underline{f}$	$\overline{FT}$	$\underline{f}$
F$\bar{T}$	$\underline{f}$	TF	tf
$\overline{FT}$	$\underline{f}$	TF	$t\underline{f}$
F$\bar{T}$	$\underline{f}$	T$\bar{F}$	$t\underline{f}$
$\overline{FT}$	$\underline{f}$	T$\bar{F}$	$t\underline{f}$
Search effort z		$\bar{z} = 1 - z$	

1.2.3 Detailed enumeration

It is worthwhile, if tedious, to list all the possible cases and to indicate to which outcomes Y_k ($k = 0, 1, 2$) they lead. To preserve generality we shall write $\alpha = \alpha_1$, $\beta = \beta_1$ for C1, with subscript 2 for C2. Moreover let

$$d_t = 1 - e^{-\alpha_1 z}, \quad d'_t = 1 - e^{-\alpha_2 \bar{z}}, \quad d_f = 1 - e^{-\beta_1 z},$$

$$d'_f = 1 - e^{-\beta_2 \bar{z}}, \quad g = \frac{\alpha_1}{\alpha_1 + \beta_1}, \quad g' = \frac{\alpha_2}{\alpha_2 + \beta_2}.$$

24   **Simple models for search in the presence of false targets**   [Ch. 1

The following 'diagrams' show all the possible cases, probabilities, and outcomes.

CELL 1 (C1)				CELL 2 (C2)			
Content	Case	Probability	Search outcome in C1	Content	Case	Probability	Search outcome in C2
TF	1	$tf[g(1-\bar{d}_t\bar{d}_f)c$	Y_1	F or F̄	(a')	fd'_fc	Y_0
TF	2	$+c\bar{c}\{g(1-\bar{d}_t\bar{d}_f)-\bar{d}_fd_t\}$	$\to 2$	F or F̄	(b')	$fd'_f\bar{c}$	Y_2
TF	3	$+c^2\{g(1-\bar{d}_t\bar{d}_f)-\bar{d}_fd_t\}$	Y_1	F or F̄	(c')	$f\bar{d}'_f$	Y_0
TF	4	$+\bar{c}\bar{d}_fd_t$	$\to 2$	F or F̄	(d')	$\bar{f}$	Y_0
TF	5	$+c^2\{\bar{g}(1-\bar{d}_t\bar{d}_f)-\bar{d}_td_f\}$	Y_1				
TF	6	$+c\bar{c}\{\bar{g}(1-\bar{d}_t\bar{d}_f)-\bar{d}_td_f\}$	$\to 2$				
TF	7	$+c\bar{d}_td_f$	$\to 2$				
TF	8	$+\bar{c}\bar{g}(1-\bar{d}_t\bar{d}_f)$	Y_1				
TF	9	$+\bar{d}_t\bar{d}_f]$	$\to 2$				
TF̄	10	$+tf[d_tc$	Y_1				
TF̄	11	$+d_t\bar{c}$	$\to 2$				
TF̄	12	$+\bar{d}_t]$	-2				

Notes:
(1) In Cell 1 (C1) cases 1 to 9 are described in words in section 1.2.2.
    Cases 10, 11, 12 correspond to T in C1 but no F. The description and outcomes are easily inferred from the probabilities.
(2) '$\to 2$' indicates that search was inconclusive in C1 and passes therefore to C2.

The next diagram is the mirror image of the first and deals with the case where C1, by protocol searched first, contains F or F̄, while T is in C2 with or without an F.

CELL 1 (C1)				CELL 2 (C2)			
Content	Case	Probability	Search outcome in C1	Content	Case	Probability	Search outcome in C2
F or F̄	a	fd_fc	$\to 2$	TF	1'	$tf[g'(1-\bar{d}'_t\bar{d}'_f)c$	Y_1
F or F̄	b	$fd_f\bar{c}$	Y_2	TF	2'	$+c\bar{c}\{g'(1-\bar{d}'_t\bar{d}'_f)-\bar{d}'_fd'_t\}$	Y_0
F or F̄	c	$f\bar{d}_f$	$\to 2$	TF	3'	$+\bar{c}^2\{g'(1-\bar{d}'_t\bar{d}'_f)-\bar{d}'_fd'_t\}$	Y_1
F or F̄	d	$\bar{f}$	$\to 2$	TF	4'	$-\bar{c}\bar{d}'_fd'_t$	Y_0
				TF	5'	$+c^2\{\bar{g}'(1-\bar{d}'_t\bar{d}'_f)-\bar{d}'_td'_f]$	Y_1
				TF	6'	$+c\bar{c}\{\bar{g}'(1-\bar{d}'_t\bar{d}'_f)-\bar{d}'_td'_f\}$	Y_0
				TF	7'	$+c\bar{d}'_td'_f$	Y_0
				TF	8'	$+\bar{c}\bar{g}'(1-\bar{d}'_t\bar{d}'_f)$	Y_1
				TF	9'	$+\bar{d}'_t\bar{d}'_f]$	Y_0
				TF̄	10'	$+\bar{t}f[d'_tc$	Y_1
				TF̄	11'	$+d'_t\bar{c}$	Y_0
				TF̄	12'	$+\bar{d}'_t]$	Y_0

The advantage of these diagrams is that it shows the construction of the probabilities and, moreover, enables simple checks to be made for consistency. Using the diagram we now list $P(Y_0)$, $P(Y_1)$, $P(Y_2)$:

$$P(Y_0) = t(1 - f\, d'_f\, \bar{c})\, B_{00} + \bar{t}(1 - f\, d_f\, \bar{c})\, B_{01}$$
$$P(Y_1) = t\, B_{10} \qquad\quad + \bar{t}(1 - f\, d_f\, \bar{c})\, B_{11}$$
$$P(Y_2) = tf\, d'_f\, c\, B_{20} + \bar{t}\, B_{21}$$

where

$$B_{00} = f\{1 - \bar{d}_t\, \bar{d}_f)\, c\bar{c} + \bar{c}^2\, \bar{d}_f\, d_t + c^2\, \bar{d}_t\, d_f + \bar{d}_t\, \bar{d}_f\} + \bar{f}(1 - d_t\, c)$$
$$B_{01} = B'_{00} \text{ (that is, } B_{00} \text{ with primable entities primed)}$$
$$B_{10} = f\{(1 - \bar{d}_t\, \bar{d}_f)(c + \bar{c}^2) - \bar{d}_f\, d_t\, \bar{c}^2 - \bar{d}_t\, d_f\, c^2\} + \bar{f}\, d_t\, c$$
$$B_{11} = B'_{10}$$
$$B_{20} = B_{00}$$
$$B_{21} = f\, d_f\, \bar{c} \qquad\qquad\qquad\qquad\qquad\qquad\qquad\qquad (37)$$

As a check we see readily that the coefficient of t in $P(Y_0) + P(Y_1) + P(Y_2)$ is:

$$(1 - f\, d'_f\, \bar{c})\, B_{00} + B_{10} + tf\, d'_f\, c\, B_{20} = B_{00} + B_{10} = f(1 - \bar{d}_t\, \bar{d}_f + \bar{d}_t\, \bar{d}_f) + \bar{f} = 1,$$

and similarly for the coefficient of $\bar{t}$.

1.2.4 Numerical calculations

A program was written to calculate $P(X_1|Y_0)$ (the posterior probability that the target is in C1, given an inconclusive outcome of the search in both cells); $P(X_1|\bar{Y}_0)$ (the posterior probability of X_1 given that the search yielded, rightly, or wrongly, a detection identified as a target); $P(Y_1)$, the probability that the conclusion drawn from the search about which cell the target is in, is correct, whether for the right or wrong reason; $P(Y_0)$, the probability that the search was inconclusive; I_E, the average increase in information yielded by the search and calculated over the set Y_0, Y_1, Y_2 of search outcomes and states X_1, X_2 of nature. This is somewhat different from calculations in Section 1.1.

To check the program and also to compare with 'Mela's first example', we give in Table 6 values for $f = 0.001$ and $c = 0.999$, that is practically no chance of false contact and almost 100 per cent assurance of correct classification. The numerical values can therefore be expected to be very close.

Although Y_0 and Y_1 have different meanings in the two models, the difference is not significant in this case: the values can be compared directly and the agreement noted.

The value of I_E reflects our choice of search outcomes.

Table 6 — Comparison with Mela's first example
Cited by Pierce [7]

z	$P(X_1|Y_0)$	$P(X_1|\overline{Y}_0)$	$P(Y_1)$	$P(Y_0)$	I_E
0.1	0.6350	0.1261	0.2653	0.7347	0.1105
0.3	0.5688	0.3281	0.2858	0.7142	0.0204
0.5	0.5	0.5	0.2926	0.7074	0
0.7	0.4312	0.6719	0.2858	0.7142	0.0240
0.9	0.3650	0.8739	0.2653	0.7347	0.1105

1.2.5 Comments

Tables 7, 8, 9 are concerned respectively with the cases $t = 0.1, 0.5$, and 0.9. *In all cases it has been assumed that* $\alpha_1 = \alpha_2 = \beta_1 = \beta_2 = 0.6931$. For the effect of different detection performance in the cells there is another set of tables available which are not given here.

The tables are set out in such a way that the effects of varying f and c are rendered as obvious as possible. Recall that t is the prior probability that the target is in C1 and that the independent variable z is the fraction of the total search effort available in C1. Let us examine briefly the columns of the tables in turn.

(1) $P(X_1|Y_0)$
  (i) This is seen to be independent of f, for given t and c. This may be something of a surprise. However, even taking different $\alpha_1, \alpha_2, \beta_1, \beta_2$ we find that

$$P(X_1|Y_0) = \frac{t(1-cd_t)}{t(1-cd_t) + \bar{t}(1-cd'_t)}$$

where $\bar{d}_t = \exp(-\alpha_1 z)$, $\bar{d}'_t = \exp(-\alpha_2(1-z))$. Thus, not only is $P(X_1|Y_0)$ independent of f, but also of β_1 and β_2 which relate to the detection of false targets.

  (ii) $P(X_1|Y_0)$ increases with c for $z < 0.5$, and decreases with c for $z > 0.5$. For $z = 0.5$ it is independent of c and indeed is equal to t. This is a special case. From the general formula we see that $P(X_1|Y_0)$ is independent of c for z such that $d_t = d'_t$, that is

$$e^{-\alpha_1 z} = e^{-\alpha_2 + \alpha_2 z},$$

that is where

$$z = \alpha_2/(\alpha_1 + \alpha_2).$$

The value of $P(X_1|Y_0)$ is then equal to the prior value t. In the particular case $\alpha_1 = \alpha_2$, $z = 0.5$.

Sec. 1.2] Two cell search: simple false target structure 27

(2) $P(X_1|\overline{Y}_0)$
  (i) As z increases this posterior increases through the prior value t for given f and c. Here the prior value occurs where $z = 0.5$.
  (ii) For fixed f it *decreases* as c increases for $z < 0.5$, and *increases* as c increases for $z > 0.5$.
  (iii) For fixed c it *increases* with f for $z < 0.5$ and decreases for $z > 0.5$.

(3) $P(Y_1)$
This is the probability that the search concludes with the correct identification of which cell the target is in. One might expect it to increase with c for fixed f and z. However, it is possible to arrive at the correct conclusions for the wrong reason, by identifying a false target as a true one. This seems more likely to happen when the false target probability is high, a tendency which might be corrected by classification capability. In fact, one observes when most effort is put into searching the cell in which the target is most likely to be a priori, that when f is high, $P(Y_1)$ decreases to a minimum and then increases again as c increases. Thus, in some conditions, to increase c does not necessarily increase $P(Y_1)$.

(4) $P(Y_0)$
  (i) This is the probability that the search ends without the target having apparently been detected. In all cases given, $P(Y_0) + P(Y_1)$ is nearly unity, and this means that $P(Y_2)$, the probability that a target is apparently identified and stated to be in the wrong cell, is rather small.
  (ii) As a function of z, $P(Y_0)$, sometimes goes to a minimum. The associated value of z may have operational importance in the design of a search.
  (iii) As a function of c, $P(Y_0)$ sometimes has a maximum.
  (iv) To increase f decreases $P(Y_0)$, as would be expected.

(5) I_E
  (i) As a function of z the average gain I_E in information yielded by the search characteristically decreases to a minimum as z increases, and then increases. It seems always least in the tables in the neighbourhood of $z = 0.5$.
  (ii) I_E increases as f increases.
  (iii) Increasing c has complex effects and needs closer investigation.

(6) Symmetry for $t = 0.5$
Not surprisingly, there are many symmetries to be seen. Note also that Table 7 and 9 present certain antisymmetries, as would be expected. For example, for $c = f = 0.1$, $P(X_1|Y_0)$ for $t = 0.1$ and given z, has the value $1 - P(X_1|Y_0)$ for $t = 0.9$ and $1 - z$. Similar observations can be made about the remaining entities.

In Section 1.3 the attempt will be made to furnish an analysis that will enable calculations similar to the above to be made for arbitrary numbers of false targets in either of the two cells.

Table 7 — $t = 0.1$

$f = 0.1$

	$c = 0.1$						$c = 0.5$						$c = 0.9$				
z	$P(X_1\|Y_0)$	$P(X_1\|\overline{Y}_0)$	$P(Y_1)$	$P(Y_0)$	I_E	$P(X_1\|Y_0)$	$P(X_1\|\overline{Y}_0)$	$P(Y_1)$	$P(Y_0)$	I_E	$P(X_1\|Y_0)$	$P(X_1\|\overline{Y}_0)$	$P(Y_1)$	$P(Y_0)$	I_E
0.1	0.1037	0.0613	0.0784	0.9120	0.0052	0.1227	0.0253	0.2278	0.7670	0.0172	0.1520	0.0169	0.3842	0.6148	0.0311
0.3	0.1018	0.0804	0.0670	0.9145	0.0015	0.1108	0.0582	0.1954	0.7944	0.0037	0.1237	0.0523	0.3301	0.6679	0.0071
0.5	0.1000	0.1000	0.0536	0.9202	0	0.1000	0.1000	0.1568	0.8287	0	0.1000	0.1000	0.2650	0.7321	0
0.7	0.0982	0.1238	0.0379	0.9294	0.0020	0.0901	0.1666	0.1111	0.8708	0.0047	0.0804	0.1828	0.1877	0.8087	0.0082
0.9	0.0964	0.1590	0.0196	0.99423	0.0100	0.0811	0.3219	0.0573	0.9216	0.0251	0.0644	0.4174	0.0966	0.0992	0.0439

$f = 0.5$

	$c = 0.1$						$c = 0.5$						$c = 0.9$				
z	$P(X_1\|Y_0)$	$P(X_1\|\overline{Y}_0)$	$P(Y_1)$	$P(Y_0)$	I_E	$P(X_1\|Y_0)$	$P(X_1\|\overline{Y}_0)$	$P(Y_1)$	$P(Y_0)$	I_E	$P(X_1\|Y_0)$	$P(X_1\|\overline{Y}_0)$	$P(Y_1)$	$P(Y_0)$	I_E
0.1	0.1037	0.8973	0.2180	0.7348	0.0294	0.1227	0.0507	0.2892	0.6847	0.0273	0.1520	0.0214	0.3931	0.6017	0.0326
0.3	0.1018	0.0950	0.1789	0.7295	0.0043	0.1108	0.0742	0.2451	0.7044	0.0046	0.1237	0.0554	0.3372	0.6526	0.0071
0.5	0.1000	0.1000	0.1386	0.7317	0.0003	0.1000	0.1000	0.1952	0.7331	0.0002	0.1000	0.1000	0.2708	0.7150	0.0001
0.7	0.0982	0.1052	0.0992	0.7414	0.0092	0.0901	0.1335	0.1379	0.7721	0.0085	0.0804	0.1738	0.1919	0.7902	0.0089
0.9	0.0964	0.1114	0.0506	0.7591	0.0370	0.0811	0.1877	0.0716	0.8228	0.0379	0.0644	0.3611	0.0989	0.8801	0.0460

$f = 0.9$

	$c = 0.1$						$c = 0.5$						$c = 0.9$				
z	$P(X_1\|Y_0)$	$P(X_1\|\overline{Y}_0)$	$P(Y_1)$	$P(Y_0)$	I_E	$P(X_1\|Y_0)$	$P(X_1\|\overline{Y}_0)$	$P(Y_1)$	$P(Y_0)$	I_E	$P(X_1\|Y_0)$	$P(X_1\|\overline{Y}_0)$	$P(Y_1)$	$P(Y_0)$	I_E
0.1	0.1037	0.0951	0.3507	0.5652	0.0489	0.1227	0.6530	0.3488	0.6045	0.0369	0.1520	0.0255	0.4020	0.5887	0.3420
0.3	0.1018	0.0976	0.2746	0.5626	0.0054	0.1108	0.0824	0.2905	0.6191	0.0051	0.1237	0.0583	0.3444	0.6375	0.0071
0.5	0.1000	0.1000	0.2041	0.5648	0.0016	0.1000	0.1000	0.2282	0.6434	0.0007	0.1000	0.1000	0.2763	0.6981	0.0104
0.7	0.0982	0.1024	0.1379	0.5718	0.0210	0.0901	0.1209	0.1600	0.6786	0.0134	0.0804	0.1663	0.1959	0.7719	0.0097
0.9	0.0964	0.1051	0.0745	0.5839	0.0703	0.0811	0.1501	0.0838	0.7264	0.0524	0.0644	0.3204	0.1011	0.8610	0.0482

Sec. 1.2] Two cell search: simple false target structure

Table 8 — $t = 0.5$

$f = 0.1$

	$c = 0.1$					$c = 0.5$					$c = 0.9$				
z	$P(X_1\|Y_0)$	$P(X_1\|\bar{Y}_0)$	$P(Y_1)$	$P(Y_0)$	I_E	$P(X_1\|Y_0)$	$P(X_1\|\bar{Y}_0)$	$P(Y_1)$	$P(Y_0)$	I_E	$P(X_1\|Y_0)$	$P(X_1\|\bar{Y}_0)$	$P(Y_1)$	$P(Y_0)$	I_E
0.1	0.5102	0.3702	0.0492	0.9272	0.0228	0.5772	0.1897	0.1429	0.8443	0.0537	0.6174	0.1343	0.2485	0.7570	0.0972
0.3	0.5051	0.4402	0.0529	0.9219	0.0044	0.5287	0.3573	0.1541	0.8326	0.0108	0.5596	0.3319	0.2592	0.7383	0.0208
0.5	0.5000	0.5000	0.0542	0.9202	0	0.5000	0.5000	0.1578	0.8287	0.0001	0.5000	0.5000	0.2654	0.7321	0
0.7	0.4949	0.5598	0.0529	0.9219	0.0053	0.4713	0.6427	0.1541	0.8326	0.0123	0.4404	0.6681	0.2592	0.7383	0.0213
0.9	0.4898	0.6298	0.0492	0.9272	0.0238	0.4428	0.8103	0.1429	0.8443	0.0554	0.3826	0.8657	0.2405	0.7570	0.0977

$f = 0.5$

	$c = 0.1$					$c = 0.5$					$c = 0.9$				
z	$P(X_1\|Y_0)$	$P(X_1\|\bar{Y}_0)$	$P(Y_1)$	$P(Y_0)$	I_E	$P(X_1\|Y_0)$	$P(X_1\|\bar{Y}_0)$	$P(Y_1)$	$P(Y_0)$	I_E	$P(X_1\|Y_0)$	$P(X_1\|\bar{Y}_0)$	$P(Y_1)$	$P(Y_0)$	I_E
0.1	0.5102	0.4699	0.1374	0.7469	0.0769	0.5572	0.3248	0.1827	0.7538	0.0782	0.6174	0.1643	0.2466	0.7409	0.1007
0.3	0.5051	0.4859	0.1446	0.7355	0.0118	0.5287	0.4191	0.1968	0.7383	0.0132	0.5596	0.3457	0.2660	0.7214	0.0208
0.5	0.5000	0.5000	0.1469	0.7317	0.0008	0.5000	0.5000	0.2013	0.7331	0.0006	0.5000	0.5000	0.2724	0.7150	0.0004
0.7	0.4949	0.5141	0.1444	0.7355	0.0251	0.4713	0.5809	0.1966	0.7383	0.0228	0.4404	0.6543	0.2660	0.7214	0.0234
0.9	0.4898	0.5301	0.1373	0.7469	0.0921	0.4428	0.6752	0.1825	0.7538	0.0894	0.3826	0.8357	0.2466	0.7409	0.1037

$f = 0.9$

	$c = 0.1$					$c = 0.5$					$c = 0.9$				
z	$P(X_1\|Y_0)$	$P(X_1\|\bar{Y}_0)$	$P(Y_1)$	$P(Y_0)$	I_E	$P(X_1\|Y_0)$	$P(X_1\|\bar{Y}_0)$	$P(Y_1)$	$P(Y_0)$	I_E	$P(X_1\|Y_0)$	$P(X_1\|\bar{Y}_0)$	$P(Y_1)$	$P(Y_0)$	I_E
0.1	0.5102	0.4862	0.2217	0.5746	0.1267	0.5572	0.3861	0.2216	0.6654	0.1017	0.6174	0.1907	0.2526	0.7248	0.1042
0.3	0.5051	0.4934	0.2272	0.5677	0.0149	0.5287	0.4470	0.2371	0.6488	0.0145	0.5596	0.3578	0.2727	0.7047	0.0208
0.5	0.5000	0.5000	0.2288	0.5648	0.0044	0.5000	0.5000	0.2419	0.6434	0.0019	0.5000	0.5000	0.2793	0.6981	0.0003
0.7	0.4949	0.5066	0.2268	0.5672	0.0567	0.4713	0.5530	0.2364	0.6488	0.0364	0.4404	0.6422	0.2726	0.7047	0.0256
0.9	0.4898	0.5138	0.2213	0.5746	0.1740	0.4428	0.6139	0.2210	0.6654	0.1269	0.3026	0.8093	0.2526	0.0256	0.1097

Table 9 — $t = 0.9$

$f = 0.1$

	$c = 0.1$					$c = 0.5$					$c = 0.9$			
z	$P(X_1\|Y_0)$	$P(X_1\|\bar{Y}_0)$	$P(Y_1)$	$P(Y_0)$	I_E	$P(X_1\|Y_0)$	$P(X_1\|\bar{Y}_0)$	$P(Y_1)$	$P(Y_0)$	I_E	$P(X_1\|Y_0)$	$P(X_1\|\bar{Y}_0)$	$P(Y_1)$	$P(Y_0)$	I_E
0.1	0.9036	0.8410	0.0200	0.9423	0.0096	0.9189	0.6781	0.0579	0.9216	0.0243	0.9356	0.5826	0.0968	0.8992	0.0436
0.3	0.9018	0.8762	0.0389	0.9294	0.0017	0.9099	0.8334	0.1127	0.8708	0.0042	0.9196	0.8172	0.1883	0.8087	0.0080
0.5	0.9000	0.9000	0.0547	0.9202	0	0.9000	0.9000	0.1587	0.8287	0	0.9000	0.9000	0.2657	0.7321	0
0.7	0.8982	0.9196	0.0679	0.9145	0.0019	0.8892	0.9418	0.1970	0.7944	0.0042	0.8763	0.8773	0.3306	0.6679	0.0073
0.9	0.8693	0.9387	0.0788	0.9120	0.0086	0.8773	0.9747	0.2285	0.7670	0.0180	0.8480	0.9831	0.3844	0.6148	0.0313

$f = 0.5$

	$c = 0.1$					$c = 0.5$					$c = 0.9$			
z	$P(X_1\|Y_0)$	$P(X_1\|\bar{Y}_0)$	$P(Y_1)$	$P(Y_0)$	I_E	$P(X_1\|Y_0)$	$P(X_1\|\bar{Y}_0)$	$P(Y_1)$	$P(Y_0)$	I_E	$P(X_1\|Y_0)$	$P(X_1\|\bar{Y}_0)$	$P(Y_1)$	$P(Y_0)$	I_E
0.1	0.9036	0.8886	0.8568	0.7591	0.0304	0.9189	0.8123	0.8763	0.8228	0.0332	0.9356	0.6389	0.1000	0.8801	0.0447
0.3	0.9018	0.8948	0.1102	0.7414	0.0043	0.9099	0.8665	0.1485	0.7721	0.0050	0.9186	0.8262	0.1946	0.7902	0.0080
0.5	0.9000	0.9000	0.1552	0.7317	0.0003	0.9000	0.9000	0.2074	0.7331	0.0002	0.9000	0.9000	0.2740	0.7150	0.0001
0.7	0.8982	0.9050	0.1927	0.7295	0.0092	0.8892	0.9258	0.2552	0.7044	0.0082	0.8763	0.9446	0.3400	0.6526	0.0081
0.9	0.8963	0.9103	0.2239	0.7348	0.0361	0.8773	0.9493	0.2934	0.6847	0.0323	0.8480	0.9786	0.3943	0.6017	0.0340

$f = 0.9$

	$c = 0.1$					$c = 0.5$					$c = 0.9$			
z	$P(X_1\|Y_0)$	$P(X_1\|\bar{Y}_0)$	$P(Y_1)$	$P(Y_0)$	I_E	$P(X_1\|Y_0)$	$P(X_1\|\bar{Y}_0)$	$P(Y_1)$	$P(Y_0)$	I_E	$P(X_1\|Y_0)$	$P(X_1\|\bar{Y}_0)$	$P(Y_1)$	$P(Y_0)$	I_E
0.1	0.9036	0.8949	0.0928	0.5839	0.0495	0.9189	0.2499	0.0944	0.7264	0.0416	0.9356	0.6796	0.1033	0.8610	0.0459
0.3	0.9018	0.8976	0.1798	0.5718	0.0054	0.9099	0.8791	0.1838	0.6786	0.0054	0.9196	0.8337	0.2009	0.7719	0.0080
0.5	0.9000	0.9000	0.2534	0.5648	0.0016	0.9000	0.9000	0.2555	0.6414	0.0007	0.9000	0.9000	0.2823	0.6981	0.0001
0.7	0.5152	0.9024	0.3156	0.5626	0.0211	0.8892	0.9176	0.3128	0.6191	0.0134	0.8763	0.0417	0.3494	0.6375	0.0090
0.9	0.8943	0.9049	0.3682	0.5652	0.0700	0.8773	0.9347	0.3581	0.6045	0.0483	0.8480	0.9745	0.4041	0.5887	0.0367

1.3 TWO CELL SEARCH: SINGLE TARGET AND ARBITRARY NUMBERS OF FALSE TARGETS

1.3.1 Basic assumptions

In this part it will be assumed that false contacts in an area constitute a spatial Poisson process whose parameter depends on the area and not on time. Writing the parameters of C1, C2, a_1, a_2 respectively, we have

$$f_n = e^{-a_1} a_1^n/n! , \quad f'_m = e^{-a_2} a_2^m/m! \ (m,n \geqslant 0) , \tag{38}$$

where f_n is the probability of n false targets in C1, f'_m that of m false targets in C2. Thus, in Section 1.2 we have put $a_1 = a_2 = a$, say, and $\bar{f} = f_0 = e^{-a}, f = 1 - e^{-a}$, where a is such that the probability of the presence of more than one false target is negligible. This can be expressed by saying that in Section 1.2 a has the property that $e^{-a}(1+a) \geqslant 0.99$, or, roughly, $a \leqslant 0.1$.

The next major step in the analysis concerns the detection process. As seen in Section 1.2, the analysis requires apparently an ordering of the detection opportunities given to the sensor by the (assumed single) target and the train of false targets existing in the particular cell under search. Thus, if the cell contains a target (T) and n false targets $F_r (r = 1, 2, \ldots, n)$, there are $n + 1$ possible orderings corresponding to placing T before F_1, after F_n, or in the $n - 1$ gaps between successive F_r. In the case of $n = 3$ we have

(1) $TF_1F_2F_3$ (2) $F_1TF_2F_3$ (3) $F_1F_2TF_3$ (4) $F_1F_2F_3T$.

The contribution of each of these configurations to the possible search outcomes Y_o, Y_1, Y_2, as defined in Section 1.2, may seem to be different.

To deal with the actual detection process it is convenient to extend the concept, also hinted at in Section 1.2, that the time from the beginning of the search required to detect an object is an exponential random variable with mean $1/\alpha$ in the case of T, and $1/\beta$ in the case of F. Let s be the time to detection of the first object encountered, $s + u_1$ to the second, ..., $s + u_1 + u_2 + \ldots + u_{n-1}$ to the nth. Then the joint probability density (JPD) of times to detection of configurations (JPD 1) to (JPD 4) above are:

$$JPD\ 1 : 3!\ \alpha e^{-\alpha s}\ \beta e^{-\beta(s+u_1)}\ \beta e^{-\beta(s+u_1+u_2)}\ \beta e^{-\beta(s+u_1+u_2+u_3)}$$

$$JPD\ 2 : 3!\ \beta e^{-\beta s}\ \alpha e^{-\alpha(s+u_1)}\ \beta e^{-\beta(s+u_1+u_2)}\ \beta e^{-\beta(s+u_1+u_2+u_3)}$$

$$JPD\ 3 : 3!\ \beta e^{-\beta s}\ \beta e^{-\beta(s+u_1)}\ \alpha e^{-\alpha(s+u_1+u_2)}\ \beta e^{-\beta(s+u_1+u_2+u_3)}$$

$$JPD\ 4 : 3!\ \beta e^{-\beta s}\ \beta e^{-\beta(s+u_1)}\ \beta e^{-\beta(s+u_1+u_1)}\ \alpha e^{-\alpha(s+u_1+u_2+u_3)}$$

The factorial is required because the F's may be ordered in 3! ways. The ranges of s, u_1, u_2, u_3 are all $(0, \infty)$ here, and to show that all possibilities have been taken into account we integrate out over these variables and show that this sum is unity.

JPD 1 gives

$$3!\,\alpha\beta^3 \int_0^\infty e^{-(\alpha+3\beta)s}\,ds \int_0^\infty e^{-3\beta u_1}\,du_1 \int_0^\infty e^{-2\beta u_2}\,du_2 \int_0^\infty e^{-\beta u_3}\,du_3 = \alpha/(\alpha+3\beta)$$

JPD 2 gives

$$3!\,\alpha\beta^3 \int_0^\infty e^{-(\alpha+3\beta)s}\,ds \int_0^\infty e^{-(\alpha+2\beta)u_1}\,du_1 \int_0^\infty e^{-2\beta u_2}\,du_2 \int_0^\infty e^{-\beta u_3}\,du_3 =$$

$$\frac{3\alpha\beta}{(\alpha+3\beta)(\alpha+2\beta)}$$

JPD 3 gives

$$3!\,\alpha\beta^3 \int_0^\infty e^{-(\alpha+3\beta)s}\,ds \int_0^\infty e^{-(\alpha+2\beta)u_1}\,du_1 \int_0^\infty e^{-(\alpha+\beta)u_2}\,du_2 \int_0^\infty e^{-\beta u_3}\,du_3 =$$

$$\frac{3!\,\alpha\beta^3}{(\alpha+3\beta)(\alpha+2\beta)(\alpha+\beta)}$$

JPD 4 gives

$$3!\,\alpha\beta^3 \int_0^\infty e^{-(\alpha+3\beta)s}\,ds \int_0^\infty e^{-(\alpha+2\beta)u_1}\,du_1 \int_0^\infty e^{-(\alpha+\beta)u_2}\,du_2 \int_0^\infty e^{-\alpha u_3}\,du_3 =$$

$$\frac{3!\,\beta^3}{(\alpha+3\beta)(\alpha+2\beta)(\alpha+\beta)}$$

Then

$$\begin{aligned}
(\text{JPD 3}) + (\text{JPD 4}) &= 3!\,\beta^2/((\alpha+3\beta)(\alpha+2\beta)) \\
(\text{JPD 2}) + (\text{JPD 3}) + (\text{JPD 4}) &= 3\beta/(\alpha+3\beta) \\
(\text{JPD 1}) + (\text{JPD 2}) + (\text{JPD 3}) + (\text{JPD 4}) &= 1\,.
\end{aligned}$$

Calculation of $P(Y_0)$, $P(Y_1)$, $P(Y_2)$

The criterion by which we determine how many targets are detected during a search of duration t (this is the 'effort', previously denoted by z) is then merely how many of the particular ordering under consideration fall within the interval $(0, t)$. If none falls in $(0, t)$, no detection can take place and search would then pass to the next cell, or a contribution to Y_0 would have been made.

It is convenient for the development of $P(Y_0)$, $P(Y_1)$, $P(Y_2)$ to consider the contributions made by search in

(i) a cell containing F's and no T

Sec. 1.3] Two cell search: single target and arbitrary numbers of false targets   33

(ii) a cell containing F's and one T

(i) Search for time t in a cell containing $F_1, F_2, \ldots, F_n$ and no T

We shall find the contribution to $P(Y_0)$. Recall that Y_0 is the outcome of a search which states that no target, apparently, has been detected. The alternative to Y_0 in a cell containing F's only is then a contribution to Y_2, the outcome of a search which states that a target is present when it really is not.

The times to detection from the beginning of search will in general be such that there exist m detection opportunities, $m = 0, 1, 2, \ldots, n$ ($n \geqslant 1$). This can be visualized as a sequence of m points F_r on a time axis to the left of the epoch t which denotes the end of the search, and $n - m$ points to the right of this epoch.

```
   F₁    F₂  ...  Fₘ        Fₘ₊₁      Fₘ₊₂  ...  Fₙ
───┼─────┼───────┼──────────┼─────────┼──────────┼─────→
   0                    t                              time
```

The probability of this distribution is

$$\binom{n}{m} \gamma^m \bar{\gamma}^{n-m}$$

where

$$\gamma = 1 - e^{-\beta t},$$
$$\bar{\gamma} = 1 - \gamma = e^{-\beta t},$$

and the contribution to $P(Y_0)$ is the above multiplied by c^m, since *all* of $F_1, F_2, \ldots, F_m$ have to be classified correctly. The order of occurrence is irrelevant. Thus, the contribution to $P(Y_0)$ for $n \geqslant 1$ is

$$\sum_{m=0}^{n} \binom{n}{m} (\gamma c)^m \bar{\gamma}^{n-m} = (\bar{\gamma} + \gamma c)^n = (1 - \gamma \bar{c})^n.$$

The interpretation of the last form is that it is the probability of the complement of the event that each F is detected and misclassified. The contribution to $P(Y_0)$ for $n = 0$ is, of course, simply the probability that no F is present. Then, deconditioning with respect to n, we have for the contribution to $P(Y_0)$

$$e^{-aC} \tag{39}$$

where

$$\overline{C} = \overline{c}\gamma. \tag{40}$$

The contribution to $P(Y_2)$ is then

$$1 - e^{-a\overline{C}}. \tag{41}$$

(ii) *Search for time t in a cell containing n F's and one T*

The argument is similar. For each disposition of $F_1, F_2, \ldots, F_n$ with respect to t there are two cases:

(a) The detection epoch of T is before t;
(b) It is after t.

Diagrammatically we have:

(a) $\quad \dfrac{F_1\ F_2\ \ldots\ T\ \ldots\ F_m \qquad F_{m+1}\ \ldots\ F_n}{0 \hspace{4cm} t \hspace{3cm} \text{time}} \longrightarrow$

(b) $\quad \dfrac{F_1\ F_2\ \ldots\ F_m \qquad F_{m+1}\ \ldots\ T\ \ldots\ F_n}{0 \hspace{3cm} t \hspace{4cm} \text{time}} \longrightarrow$

and, of course, T may lie before or after the sequence of F's, or may lie between two of them. In either case the overall probabilities of the dispositions are

(a) $\quad \delta \dbinom{n}{m} \gamma^m \overline{\gamma}^{n-m} \qquad (\delta = 1 - e^{-\alpha t}) \tag{42}$

(b) $\quad \overline{\delta} \dbinom{n}{m} \gamma^m \overline{\gamma}^{n-m}. \tag{43}$

The contribution to $P(Y_0)$ in this case is that all detected F's are correctly classified and that T is incorrectly classified. Moreover, in this case, the search continues until all detectable objects have been classified. Thus, conditionally, we have from (a) the contribution

$$\delta \binom{n}{m} \gamma^m \overline{\gamma}^{n-m} c^m \overline{c}, \tag{44}$$

and from (b)

$$\overline{\delta} \binom{n}{m} \gamma^m \overline{\gamma}^{n-m} c^m. \tag{45}$$

Sec. 1.3] Two cell search: single target and arbitrary numbers of false targets

It is readily verified that these results are independent of the order of occurrence of the contacts. Hence, for $n \geq 1$, (a) and (b) give the following contributions to $P(Y_0)$.

(a) $\quad \bar{c}\bar{\delta}(c\gamma + \gamma)^n = \bar{c}\bar{\delta}\, C^n \quad$ (46)

(b) $\quad \bar{\delta}\, C^n$. $\quad$ (47)

This is for $n \geq 1$, but clearly holds also for $n = 0$. Deconditioning we get the following contribution to $P(Y_0)$:

$$De^{-aC} \quad (48)$$

where

$$D = 1 - c\delta \quad (49)$$

is the probability of the complement of the event that T is detected and correctly classified. Naturally we also write

$$\bar{D} = c\delta. \quad (50)$$

We can now construct $P(Y_0)$ for a two-cell search using the protocol of Section 1.2 for search in cell 1 first. Let p, q be respectively the probabilities that T lies in C1, C2. Then

$$\begin{aligned} P(Y_0) &= pD_1 \exp(-a_1\bar{C}_1 - a_2\bar{C}_2) + qD_2 \exp(-a_1\bar{C}_1 - a_2\bar{C}_2) = \\ &= (pD_1 + qD_2) \exp(-a_1\bar{C}_1 - a_2\bar{C}_2) \end{aligned} \quad (51)$$

where $\alpha_1, \beta_1, \alpha_2, \beta_2$ are the detection parameters in C_1, C_2; a_1, a_2 are the parameters of the false target stream, $\bar{t} = 1 - t$, and

$$\bar{C}_1 = c(1 - e^{-\beta_1 t}) = \bar{c}\gamma_1, \quad \bar{C}_2 = \bar{c}\gamma_2, \quad (52)$$

$$\bar{D}_1 = c(1 - e^{-\alpha_1 t}) = c\delta_1, \quad \bar{D}_2 = c\delta_2. \quad (53)$$

Note that the search time in C_2 is $\bar{t} = 1 - t$.

Next we find $P(Y_2)$. If T is in C1 the contribution of C1 is the same as that to $P(Y_0)$, while the contribution of C2 is given by (41); if T is in C2, search in C1 must give an apparent target detection although C1 contains nothing, or F's. This is again the complement of Y_0. Thus

$$P(Y_2) = pD_1 \exp(-a_1\bar{C}_1)(1 - \exp(-a_2\bar{C}_2)) + q(1 - \exp(-a_1\bar{C}_1)). \quad (54)$$

Using

$$P(Y_0) + P(Y_1) + P(Y_2) = 1 \tag{55}$$

we deduce that

$$P(Y_1) = p(1 - D_1 \exp(-a_1\overline{C}_1)) + q\exp(-a_1\overline{C}_1)(1 - D_2 \exp(-a_2\overline{C}_2)). \tag{56}$$

It will be noticed that $P(X_1|Y_0)$ is independent of the 'false target rate', as observed in Section 1.2.

An independent verification of (56) is of interest. The contributions of (a) and (b) are $(1 - c^m \overline{c})$ and $(1 - c^m)$, whatever the order of the F's and T. Thus the overall contributions of (a) and (b) for given n are the sum over m from zero to n of

(a) $\quad \delta \binom{n}{m} \gamma^m \overline{\gamma}^{n-m} (1 - c^m \overline{c})$

and

(b) $\quad \overline{\delta} \binom{n}{m} \gamma^m \overline{\gamma}^{n-m} (1 - c^m)$

giving

(a) $\quad \delta(1 - \overline{c}(\overline{\gamma} + c\gamma)^n) = \delta(1 - \overline{c}\, C^n)$

and

(b) $\quad e^{-\alpha t}(1 - C^n)$.

Combining these gives

$$\delta(1 - \overline{c}\exp(-a\overline{C})) + \overline{\delta}(1 - \exp(-a\overline{C})) = 1 - D\exp(-a\overline{C}).$$

Thus, putting it into the actual two cell search context we get

$$P(Y_1) = p(1 - D_1 \exp(-a_1\overline{C}_1)) + q\exp(-a_1\overline{C}_1)(1 - D_2 \exp(-a_2\overline{C}_2)).$$

In this form the problem of allocation of search effort t in such a way as to maximize $P(Y_1)$, for example, seems possibly capable of an analytic approach. Leaving this on one side for a future instalment we next examine the results of some calculations, primarily to determine to what extent this generalization of the model by comparison with Section 1.2 has changed the picture.

Sec. 1.4] Two cell search: two targets and arbitrary numbers of false targets

The numerical values given in Tables 10, 11, 12 gave been calculated with the intention to facilitate comparison with Section 1.2. The program requires values of the following parameters: p (previously t), $a_1, a_2, c, \alpha_1, \alpha_2, \beta_1, \beta_2$. In what follows, $\alpha_1 = \alpha_2 = \beta_1 = \beta_2 = 0.6931$, as in the sequence of Tables 7, 8, 9 of Section 1.2.

To compare the output of the model with that of Section 1.2, first we put $a_1 = a_2 = 0.001$. This means that the probability of no F in a cell is $\exp(-0.001) = 0.999$. This is $\bar{f}$ of the earlier parts and, in particular, enables a table to be generated to compare with Table 6 of Section 1.2. The agreement is exact. c is, of course, 0.999.

We turn now to the value $f = 0.1$ in Section 1.2. This means that $\bar{f} = 0.9 = e^{-a}$, giving $a = 0.1054$. It is therefore a comfort to find that values for $a = 0.1054$ are very close to those given by the program used to compute Tables 7 to 9 of Section 1.2.

To obtain values to compare with $f = 0.5, 0.9$ of Section 1.2 we have chosen a such that $e^{-a} = \bar{f}$. This gives $a = 0.6931, 2.3026$ respectively.

We make only a few comments on the tables since the pattern is similar to that shown in Section 1.2 with the changes to be expected by an allowance for more than one false target in the cells.

(1) $P(X_1|Y_0)$
This is independent of a and has exactly the same value as in Section 1.2.

(2) $P(X_1|\bar{Y}_0)$
This is less than in Section 1.2 for small c. Otherwise it shows less variation with respect to t. The most remarkable difference is that it does not pass through the prior value when $t = 0.5$, but a value rather less.

(3) $P(Y_1)$
This is greater than in Section 1.2, one of the results of higher false contact incidence and hence higher chance of mistaken identity leading to the identification of an object as a target when it is not. At high false target levels $P(Y_1)$ decreases as c increases, the increased capability decreasing the number of misidentifications, a tendency noted also in Section 1.2.

(4) $P(Y_0)$
The values tend to be less than in Section 1.2.

(5) I_E
The average gain is again minimized with respect to effort near $z = 0.5$ for all p (previously denoted by t) considered.

1.4 TWO CELL SEARCH: TWO TARGETS AND ARBITRARY NUMBERS OF FALSE TARGETS

1.4.1 Introduction

It is interesting as well as important to assess the effect of the presence of additional genuine targets. To generalize Section 1.3 completely in the same detail seems at

Table 10 — $p = 0.1$

$a = 0.6931$ $(\bar{f} = 0.5)$

	$c = 0.1$					$c = 0.5$					$c = 0.9$			
z	$P(X_1\|Y_0)$	$P(X_1\|\bar{Y}_0)$	$P(Y_1)$	$P(Y_0)$	I_E	$P(X_1\|Y_0)$	$P(X_1\|\bar{Y}_0)$	$P(Y_1)$	$P(Y_0)$	I_E	$P(X_1\|Y_0)$	$P(X_1\|\bar{Y}_0)$	$P(Y_1)$	$P(Y_0)$	I_E
0.1	0.1037	0.0888	0.2622	0.6875	0.0327	0.1227	0.0557	0.3171	0.6554	0.0305	0.1520	0.0252	0.3990	0.5967	0.0335
0.2	0.1028	0.0890	0.2482	0.6793	0.0141	0.1166	0.0651	0.1993	0.6617	0.0141	0.1371	0.0390	0.3733	0.6193	0.0173
0.3	0.1018	0.0895	0.2318	0.6743	5E–03	0.1108	0.0742	0.2788	0.6706	5.3E–03	0.1237	0.0561	0.3448	0.6456	7.3E–03
0.4	0.1009	0.0905	0.2130	0.6724	9E–04	0.1053	0.0836	0.2554	0.6822	1.1E–03	0.1113	0.0754	0.3132	0.6748	7.3E–03
0.5	0.1000	0.0919	0.1914	0.6736	3E–04	0.1000	0.0938	0.2288	0.6967	2E–04	0.1000	0.0983	0.2784	0.7071	1E–04
0.6	0.0991	0.0938	0.1668	0.6779	2.7E–03	0.0949	0.1055	0.1989	0.7143	2.5E–03	0.0897	0.1273	0.2401	0.7427	2.2E–03
0.7	0.0982	0.0965	0.1273	0.6853	8.4E–03	0.0901	0.1197	0.1651	0.7351	8.2E–03	0.0720	0.1273	0.1521	0.8245	0.0217
0.9	0.0964	0.1050	0.0712	0.7104	0.0374	0.0811	0.1642	0.0847	0.7875	0.0386	0.0644	0.3361	0.1018	0.8713	0.0461

$a = 2.3026$ $(\bar{f} = 0.1)$

	$c = 0.1$					$c = 0.5$					$c = 0.9$				
z	$P(X_1\|Y_0)$	$P(X_1\|\bar{Y}_0)$	$P(Y_1)$	$P(Y_0)$	I_E	$P(X_1\|Y_0)$	$P(X_1\|\bar{Y}_0)$	$P(Y_1)$	$P(Y_0)$	I_E	$P(X_1\|Y_0)$	$P(X_1\|\bar{Y}_0)$	$P(Y_1)$	$P(Y_0)$	I_E
0.1	0.1037	0.0886	0.5855	0.3186	0.0565	0.1227	0.0781	0.5054	0.4275	0.0511	0.1520	0.3660	0.4365	0.5169	0.0393
0.3	0.1028	0.0835	0.5678	0.3040	0.0190	0.1166	0.0790	0.4854	0.4233	0.1990	0.1373	0.0500	0.4108	0.4664	0.00192
0.4	0.1009	0.0790	0.5114	0.2893	2.4E–03	0.1053	0.0823	0.4287	0.4270	1.4E–03	0.1113	0.0784	0.3487	0.6145	0.1.8E–03
0.5	0.1000	0.7870	0.4709	0.2883	5.6E–03	0.1000	0.0849	0.3907	0.4349	2.3E–03	0.1000	0.0951	0.3116	0.6435	1.8E–03
0.6	0.0991	0.0795	0.4206	0.2916	0.0134	0.0949	0.0885	0.3453	0.4470	7.8E–03	0.0897	0.1152	0.2701	0.6762	3.1E–03
0.7	0.0982	0.0815	0.3582	0.2992	0.0261	0.0901	0.0934	0.2911	0.4639	0.0181	0.0804	0.1410	0.2238	0.7129	0.0108
0.8	0.0973	0.0852	0.0281	0.3115	0.0461	0.0855	0.1005	0.2266	0.4858	0.0354	0.0720	0.1772	0.1723	0.7540	0.0251
0.9	0.0964	0.0915	0.1854	0.3291	0.0822	0.0811	0.1110	0.1498	0.5136	0.6710	0.0644	0.2343	0.1149	0.7999	0.0521

Sec. 1.4] **Two cell search: two targets and arbitrary numbers of false targets** 39

Table 11 — $p = 0.5$

$a = 0.6931$ $(\bar{f} = 0.5)$

	c = 0.1					c = 0.5					c = 0.9				
z	$P(X_1\|Y_0)$	$P(X_1\|\bar{Y}_0)$	$P(Y_1)$	$P(Y_0)$	I_E	$P(X_1\|Y_0)$	$P(X_1\|\bar{Y}_0)$	$P(Y_1)$	$P(Y_0)$	I_E	$P(X_1\|Y_0)$	$P(X_1\|\bar{Y}_0)$	$P(Y_1)$	$P(Y_0)$	I_E
0.1	0.5102	0.4672	0.1667	0.6989	0.8650	0.5572	0.3468	0.2009	0.7215	0.0866	0.6174	0.1761	0.2504	0.7335	0.1028
0.2	0.5076	0.4678	0.1777	0.6877	0.0392	0.5430	0.3851	0.2133	0.7105	0.0407	0.5888	0.2677	0.2627	0.7219	0.0518
0.3	0.5051	0.4695	0.1853	0.6798	0.0142	0.5287	0.4191	0.2219	0.7028	0.0155	0.5596	0.3486	0.2714	0.7137	0.0214
0.4	0.5025	0.4725	0.1899	0.6752	2.6E-03	0.5144	0.4510	0.2271	0.6983	3.2E-03	0.5299	0.4233	0.2767	0.7087	5.1E-03
0.5	0.5000	0.4767	0.1914	0.6736	8E-04	0.5000	0.4824	0.2288	0.6967	7E-04	0.5000	0.4953	0.2784	0.7071	2E-04
0.6	0.4975	0.4824	0.1899	0.6752	7–8E-03	0.4856	0.5150	0.2271	0.6983	7E-03	0.4701	0.5675	0.2767	0.7087	6.1E-03
0.7	0.4949	0.4900	0.1853	0.6798	0.0239	0.4713	0.5504	0.2219	0.7028	0.0226	0.4404	0.6432	0.2714	0.7137	0.0234
0.8	0.4924	0.5001	0.1777	0.6877	0.0517	0.4570	0.5907	0.2133	0.7105	0.0497	0.4112	0.7259	0.2627	0.7219	0.0543
0.9	0.4898	0.5135	0.1667	0.6989	0.0980	0.4428	0.6388	0.2009	0.7215	0.0948	0.3826	0.8201	0.2504	0.7335	0.1049

$a = 2.3026$ $(\bar{f} = 0.1)$

	c = 0.1					c = 0.5					c = 0.9				
z	$P(X_1\|Y_0)$	$P(X_1\|\bar{Y}_0)$	$P(Y_1)$	$P(Y_0)$	I_E	$P(X_1\|Y_0)$	$P(X_1\|\bar{Y}_0)$	$P(Y_1)$	$P(Y_0)$	I_E	$P(X_1\|Y_0)$	$P(X_1\|\bar{Y}_0)$	$P(Y_1)$	$P(Y_0)$	I_E
0.1	0.5102	0.4667	0.3855	0.3239	0.1543	0.5572	0.4325	0.3276	0.4706	0.1397	0.6174	0.2550	0.2757	0.6734	0.1162
0.2	0.5076	0.4504	0.4244	0.3078	0.0552	0.5430	0.4356	0.3560	0.4546	0.1024	0.5888	0.3213	0.2916	0.6602	0.0566
0.3	0.5051	0.4407	0.4508	0.2968	0.0801	0.5287	0.4402	0.3755	0.4435	0.0182	0.5596	0.3797	0.3027	0.6509	0.0224
0.4	0.5025	0.4357	0.4660	0.2904	6.8E-03	0.5144	0.4465	0.3870	0.4370	4.1E-03	0.5299	0.4337	0.3094	0.6453	5.1E-03
0.5	0.5000	0.4346	0.4709	0.2883	0.0168	0.5000	0.4550	0.3907	0.4349	6.9E-03	0.5000	0.4861	0.3116	0.6435	9E-04
0.6	0.4975	0.4372	0.4660	0.2904	0.0412	0.4856	0.4662	0.3870	0.4370	0.0234	0.4701	0.5395	0.3094	0.6453	8.9E-03
0.7	0.4949	0.4440	0.4508	0.2968	0.0801	0.4713	0.4811	0.3755	0.4435	0.0539	0.4404	0.5963	0.3027	0.6509	0.2950
0.8	0.4924	0.4581	0.4244	0.3078	0.1387	0.4570	0.5013	0.3560	0.4546	0.1024	0.4112	0.6597	0.2916	0.6602	0.0654
0.9	0.4898	0.4755	0.3855	0.3239	0.2334	0.4428	0.5291	0.3276	0.4706	0.1815	0.3826	0.7336	0.2757	0.6734	0.1241

Table 12 — $p = 0.9$

$a = 0.6931$	($\bar{f} = 0.5$)														
		$c = 0.1$					$c = 0.5$					$c = 0.9$			
z	$P(X_1\|Y_0)$	$P(X_1\|\bar{Y}_0)$	$P(Y_1)$	$P(Y_0)$	I_E	$P(X_1\|Y_0)$	$P(X_1\|\bar{Y}_0)$	$P(Y_1)$	$P(Y_0)$	I_E	$P(X_1\|Y_0)$	$P(X_1\|\bar{Y}_0)$	$P(Y_1)$	$P(Y_0)$	I_E
0.1	0.9036	0.8876	0.0712	0.7104	0.0342	0.9189	0.8269	0.0847	0.7875	0.0364	0.9356	0.6580	0.1018	0.8713	0.0455
0.2	0.9027	0.8878	0.1071	0.6961	0.0149	0.9189	0.8269	0.0847	0.7875	0.0364	0.9280	0.7669	0.1520	0.0245	0.0210
0.3	0.9018	0.8885	0.1388	0.6853	5.3E–03	0.9099	0.8666	0.1651	0.7351	5.9E–03	0.9196	0.8281	0.1981	0.7817	8.3E–03
0.4	0.9009	0.8896	0.1668	0.6779	1E–03	0.9051	0.8809	0.1989	0.7143	1.2E–03	0.9103	0.8685	0.2401	0.7427	1.9E–03
0.5	0.9000	0.8913	0.1914	0.6736	3E–04	0.9000	0.8935	0.2288	0.6967	3E–04	0.9000	0.8983	0.2784	0.7071	1E–04
0.6	0.8991	0.8935	0.2130	0.6724	3E–04	0.8947	0.9053	0.2554	0.6822	2.6E–03	0.8887	0.9219	0.3132	0.6748	2.2E–03
0.7	0.8982	0.8963	0.2318	0.6743	9.1E–03	0.8892	0.9168	0.2788	0.6706	8.4E–03	0.8763	0.9419	0.3448	0.6456	8.2E–03
0.8	0.8972	0.9000	0.2482	0.6793	0.0201	0.8834	0.9285	0.2993	0.6617	0.0185	0.8627	0.9597	0.3773	0.6193	0.0185
0.9	0.8963	0.9048	0.2622	0.6875	0.0394	0.8773	0.9409	0.3171	0.6554	0.0354	0.8480	0.9763	0.3990	0.5957	0.0348

$a = 2.3026$	($\bar{f} = 0.1$)														
		$c = 0.1$					$c = 0.5$					$c = 0.9$			
z	$P(X_1\|Y_0)$	$P(X_1\|\bar{Y}_0)$	$P(Y_1)$	$P(Y_0)$	I_E	$P(X_1\|Y_0)$	$P(X_1\|\bar{Y}_0)$	$P(Y_1)$	$P(Y_0)$	I_E	$P(X_1\|Y_0)$	$P(X_1\|\bar{Y}_0)$	$P(Y_1)$	$P(Y_0)$	I_E
0.1	0.9036	0.8873	0.1854	0.3291	0.0605	0.9189	0.8728	0.1497	0.5136	0.5570	0.9356	0.7550	0.1149	0.7999	0.5000
0.2	0.9027	0.8806	0.2810	0.3115	0.0214	0.9145	0.8741	0.2266	0.4858	0.0221	0.9280	0.8099	0.1723	0.7540	0.0226
0.3	0.9018	0.8764	0.3582	0.2992	6.1E–03	0.9099	0.8762	0.2911	0.4639	7E–03	0.9196	0.8463	0.2238	0.7129	8.6E–03
0.4	0.9009	0.8742	0.4206	0.2916	2.5E–03	0.9051	0.8789	0.3453	0.4470	1.5E–03	0.9103	0.8733	0.2701	0.6762	1.9E–03
0.5	0.9000	0.8737	0.4709	0.2883	6.6E–03	0.9000	0.8825	0.3907	0.4349	2.6E–03	0.9000	0.8949	0.3116	0.6435	4E–04
0.6	0.8991	0.8749	0.4709	0.2083	6.6E–03	0.8947	0.8871	0.4287	0.4270	9.3E–03	0.8887	0.9134	0.3487	0.6145	8.3E–03
0.7	0.8982	0.8779	0.5433	0.2944	0.0338	0.8892	0.8930	0.4600	0.4231	0.0219	0.8763	0.9300	0.3816	0.5888	0.0108
0.8	0.8972	0.8830	0.5678	0.3040	0.0599	0.8834	0.9005	0.4854	0.4233	0.0423	0.8627	0.9451	0.4108	0.5664	0.0236
0.9	0.8963	0.8908	0.5855	0.3186	0.1033	0.8773	0.9100	0.5054	0.4275	0.0762	0.8480	0.9612	0.4365	0.5469	0.0442

Sec. 1.4] **Two cell search: two targets and arbitrary numbers of false targets** 41

present to be quite complicated, but it is manageable for the case of two targets, and the details and results are worth recording for the illumination shed on the general problems of search and information. However, the increased complexity of specifying an appropriate inference space is illustrated, together with the need for care in doing so: it is possible to specify a space which makes the prospect of the searcher quite dim, and another which gives a distinct basis for optimism.

1.4.2 The model

The notation of Section 1.3 is continued. For completeness we first recall the definition of the frequently recurring symbols. The first cell searched is referred to as C1. It is given in this case that two targets are present and the possible target distribution is accordingly.

State of nature	C1	C2	With probability
X_1	$T_1 T_2$	—	p^2
X_2	T_1	T_2	$2pq$ (the order is immaterial)
X_3	—	$T_1 T_2$	q^2

If the search outcome is two or more detections in C1 classified as target (though they may not be), the search stops: otherwise it continues into C2.

The time (effort) allocated to C1 is t: that to C2 is $\bar{t} = 1 - t$.

False targets are assumed to occur in a spatial Poisson stream independently in C1 and C2 with parameters a_1, a_2 respectively. Given that a cell contains n false targets, a search for finite time t will provide a random number of false target detection opportunities, a detection opportunity being an 'encounter' with the false target before search time in the cell is exhausted. The probability that a false target detection opportunity occurs in a search of duration t is $\gamma = 1 - e^{-\beta t}$, and we write $\bar{\gamma} = 1 - \gamma = e^{-\beta t}$. By a well-known property of the Poisson distribution the conditional probability of there being m false target detection opportunities in a search of duration t is given by

$$\binom{n}{m} \gamma^m \bar{\gamma}^{n-m} .$$

For generality a distinction is made between C1 and C2 by providing for different values of the parameters $\beta, \beta_1,$ and β_2. Note that whereas $\bar{\gamma}_1 = \exp(-\beta_1 t), \bar{\gamma}_2 = \exp(-\beta_2 \bar{t})$.

A similar argument applies to genuine target detection opportunities. A cell containing two targets provides two detection opportunities in a search of duration t with probability δ^2, one with probability $2\delta\bar{\delta}$, and none with probability $\bar{\delta}^2$, where $\delta = 1 - e^{-\alpha t}, \bar{\delta} = 1 - \delta$. The values of α and t for C1 and C2 are respectively $\alpha_1, t, \alpha_2 \bar{t}$.

The analysis also incorporates an assumed binary classification capability. With probability c an object is correctly identified: with probability $\bar{c}$ it is identified as the complement of what it is.

There are six mutually exclusive and exhaustive outcomes of the two cell search. These are illustrated below: the numbers are those of detections classified rightly or wrongly as 'target'.

Outcome	C1	C2
(1)	$\geq 2^\dagger$	—
(2)	1	≥ 1
(3)	1	0
(4)	0	0
(5)	0	1
(6)	0	$\geq 2^\dagger$

† If there were an excess over 2 it would naturally be assumed that false targets were the cause. By virtue of there having occurred two or more apparent target detections in C1 the assumption is that search would not continue into C2, it being given that the total number of targets available is two.

We now list the probabilities of (1) to (6) corresponding to the three states of nature X_1, X_2, X_3 enumerated at the start of this section.

The diagram shows the three possible occurrences in time of the detection opportunities in a search of duration t with m out of n false target detection opportunities.

$$F_1 \ldots F_m \overset{t}{\downarrow} F_{m+1} \ldots F_n$$

$$\overset{\uparrow}{t}$$

(a) $T_1 \quad T_2$
(b) $T_1 \qquad\qquad T_2$
(c) $— \qquad T_1 \quad T_2$

The conditional probability of (1) is accordingly:

$$\delta_1^2\{c^2 + (2c\bar{c})(1-c^m) + \bar{c}^2(1-c^m - mc^{m-1}\bar{c})\} + \\ + 2\delta_1\bar{\delta}_1\{c(1-c^m) + \bar{c}(1-c^m - mc^{m-1}\bar{c})\} + \bar{\delta}_1^2\{1-c^m - mc^{m-1}\bar{c}\} .$$

The notation makes the argument transparent. Note that the order in which the opportunities occur is discounted. It is as though in case (a) 2 T's and m F's are placed in a box, picked out one by one in any order, examined, and identified correctly with probability c, incorrectly with probability $\bar{c}$. The expression can be reduced to

Sec. 1.4] Two cell search: two targets and arbitrary numbers of false targets

$$\delta_1^2\{1 - c^m(1-c^2) - \bar{c}^2 \, m \, c^{m-1}\bar{c}\} + 2\delta_1 \bar{\delta}_1 (1 - c^m - \bar{c} \, mc^{m-1}\bar{c}) + \bar{\delta}_1(1 - c^m - mc^{m-1}\bar{c})$$

$$= 1 - c^m\{\delta_1^2(1-c^2) + 2\delta_1 \bar{\delta}_1 + \bar{\delta}_1^2\} - (\bar{\delta}_1 + \bar{c}\,\delta_1)^2 \, mc^{m-1}\bar{c}$$

$$= 1 - c^m(1 - \delta_1^2 c^2) - (1 - \delta_1 c)^2 \, mc^{m-1}\bar{c} \; .$$

We vary the notation of Section 1.3 to avoid confusion of c and C. Thus we shall write

$$c\delta = \bar{D}, \qquad \bar{c}\gamma = \bar{\Gamma}$$

with subscripts according to the cell. The expression can then be written

$$1 - c^m(1 - \bar{D}_1^2) - D_1^2 \, mc^{m-1}\bar{c} \; .$$

Recalling that

$$\sum_{m=0}^{n} \binom{n}{m} \gamma^m \, \bar{\gamma}^{n-m} c^m = (\bar{\gamma} + \gamma c)^n = \Gamma^n$$

and

$$\sum_{m=0}^{n} \binom{n}{m} mc^{m-1} \gamma^m \, \bar{\gamma}^{n-m} = n\gamma \, \Gamma^n \; ,$$

and deconditioning the expression with respect to m gives

$$1 - (1 - \bar{D}_1^2) \Gamma_1^n - D_1^2 \, \bar{c} \, n\gamma \, \Gamma_1^{n-1} = 1 - (1 - \bar{D}_1^2) \Gamma_1^n - D_1^2 \, n \, \bar{\Gamma}_1 \, \Gamma_1^{n-1} \; .$$

Finally, deconditioning with respect to n gives for the probability $P((1)|X_1)$ of (1), given that the state of nature is X_1,

$$P((1)|X_1) = e^{-a_1} \sum_{n \geq 0} \frac{(a_1)^n}{n!} \{1 - (1 - \bar{D}_1^2)\Gamma_1^n - D_1^2 \, \bar{\Gamma}_1^n \, \Gamma_1^n\}$$

$$= 1 - (1 - \bar{D}_1^2) \exp(-a_1\bar{\Gamma}_1) - D_1^2 \bar{\Gamma}_1 a_1 \exp(-a_1\bar{\Gamma}_1) \; .$$

It is not necessary to describe the remaining cases in such detail. We list below all eighteen $P((i)|X_j)$ for $i = 1, 2, \ldots 6, j = 1, 2, 3$.

State of nature X_1

$$P((1)|X_1) = 1 - (1 - \overline{D}_1^2) \exp(-a_1\overline{\Gamma}_1) - D_1^2 \, \overline{\Gamma}_1 \, a_1 \exp(-a_1\overline{\Gamma}_1) \qquad (57)$$

$$P((2)|X_1) = D_1 \exp(-a_1\overline{\Gamma}_1)(1 - \exp(-a_2\overline{\Gamma}_2))(2\overline{D}_1 + a_1 \, \overline{\Gamma}_1 \, D_1) \qquad (58)$$

$$P((3)|X_1) = D_1 \exp(-a_1\overline{\Gamma}_1 - a_2\overline{\Gamma}_2)(2\overline{D}_1 + a_1 \, \overline{\Gamma}_1 \, D_1) \qquad (59)$$

$$P((4)|X_1) = D_1^2 \exp(-a_1\overline{\Gamma}_1 - a_2\overline{\Gamma}_2) \qquad (60)$$

$$P((5)|X_1) = D_1^2 \, \overline{\Gamma}_2 \, a_2 \exp(-a_1\overline{\Gamma}_1 - a_2\overline{\Gamma}_2) \qquad (61)$$

$$P((6)|X_1) = D_1^2 \exp(-a_1\overline{\Gamma}_1)(1 - \exp(-a_2\overline{\Gamma}_2) - a_2 \, \overline{\Gamma}_2 \exp(-a_2\overline{\Gamma}_2)) \, . \qquad (62)$$

State of nature X_2

$$P((1)|X_2) = 1 - \exp(-a_1\overline{\Gamma}_1) - D_1 \, \overline{\Gamma}_1 \, a_1 \exp(-a_1\overline{\Gamma}_1) \qquad (63)$$

$$P((2)|X_2) = \exp(-a_1\overline{\Gamma}_1)(\overline{D}_1 + D_1\overline{\Gamma}_1 a_1)(1 - D_2 \exp(-a_2\overline{\Gamma}_2)) \qquad (64)$$

$$P((3)|X_2) = D_2 \exp(-a_1\overline{\Gamma}_1 - a_2\overline{\Gamma}_2)(\overline{D}_1 + D_1\overline{\Gamma}_1 a_1) \qquad (65)$$

$$P((4)|X_2) = D_1 \, D_2 \exp(-a_1\overline{\Gamma}_1 - a_2\overline{\Gamma}_2) \qquad (66)$$

$$P((5)|X_2) = D_1 \exp(-a_1\overline{\Gamma}_1 - a_2\overline{\Gamma}_2)(\overline{D}_2 + a_2 \, D_2 \, \overline{\Gamma}_2) \, . \qquad (67)$$

$$P((6)|X_2) = D_1 \exp(-a_1\overline{\Gamma}_1)(1 - \exp(-a_2\overline{\Gamma}_2) - D_2 \, \overline{\Gamma}_2 \, a_2 \exp(-a_2\overline{\Gamma}_2)) \qquad (68)$$

State of nature X_3

$$P((1)|X_3) = 1 - \exp(-a_1\overline{\Gamma}_1) - a_1 \, \overline{\Gamma}_1 \exp(-a_1\overline{\Gamma}_1) \qquad (69)$$

$$P((2)|X_3) = a_1 \, \overline{\Gamma}_1 \exp(-a_1\overline{\Gamma}_1)(1 - D_2^2 \exp(-a_2\overline{\Gamma}_2)) \qquad (70)$$

$$P((3)|X_3) = a_1 \, \overline{\Gamma}_1 \, D_2^2 \exp(-a_1\overline{\Gamma}_1 - a_2\overline{\Gamma}_2) \qquad (71)$$

$$P((4)|X_3) = D_2^2 \exp(-a_1\overline{\Gamma}_1 - a_2\overline{\Gamma}_2) \qquad (72)$$

$$P((5)|X_3) = D_2 \exp(-a_1\overline{\Gamma}_1 - a_2\overline{\Gamma}_2)(2\overline{D}_2 + a_2 \, \overline{\Gamma}_2 \, D_2) \qquad (73)$$

$$P((6)|X_3) = \exp(-a_1\overline{\Gamma}_1)\{1 - (1 - \overline{D}_2^2) \exp(-a_2\overline{\Gamma}_2) - D_2^2 \, \overline{\Gamma}_2 \, a_2 \exp(-a_2\overline{\Gamma}_2)\} \, . \qquad (74)$$

It is of interest to identify the probabilistic meanings of the components of these expressions, but we leave this to the reader.

An important check is to verify that

$$\sum_{i=1}^{6} P((i)|X_j) = 1 \qquad (j = 1, 2, 3) \, , \qquad (75)$$

which has been done.

Sec. 1.4] **Two cell search: two targets and arbitrary numbers of false targets** 45

1.4.3 Decision state space
The next step is to define a suitable decision state space. We consider two possibilities.

(i) Y_0 Search inconclusive because either no, or at most one, target-like identifications were made when it is known that two targets are present.
Y_1 At least two target-like identifications altogether, and in the correct cell.
Y_2 At least two target-like identifications altogether in the wrong cell, or with the wrong distribution.

(ii) Z_0 No target-like identification in either cell.
Z_1 At least one target-like identification in a cell where there is a target.
Z_2 No target-like identification in a cell where there is a target.

Table 13 shows how the six possible search outcomes contribute to these spaces.

Table 13 — Contributions of all combinations of observed and true states in Y and Z spaces

Observed state			True state					
			X_1		X_2		X_3	
Number	Content of C1	C2	C1	2	1		0	
			C2	0	1		2	
			Y	Z	Y	Z	Y	Z
(1)	≥ 2	—	Y_1	Z_1	Y_2	Z_1	Y_2	Z_2
(2)	1	≥ 1	Y_2	Z_1	Y_1	Z_1	Y_2	Z_1
(3)	1	0	Y_0	Z_1	Y_0	Z_1	Y_0	Z_2
(4)	0	0	Y_0	Z_0	Y_0	Z_0	Y_0	Z_1
(5)	0	1	Y_0	Z_2	Y_0	Z_1	Y_0	Z_1
(6)	0	≥ 2	Y_2	Z_2	Y_2	Z_1	Y_1	Z_1

The set Y_i ($i = 0, 1, 2$) was defined as a direct extension of the set Y_i defined in Section 1.3 in the case of a single target. In that case Y_0 meant 'an inconclusive search', and its extension to two targets is unexceptionable. Y_1 meant 'a target-like detection in the cell where the target is': the extension to two targets seems more restrictive since it requires not only the identification of the number of targets but also the correct distribution. Y_2 meant 'a target-like detection but in the wrong cell'. In the two-target case Y_2 arises also because of incorrect observation of the distribution.

The set Z_i ($i = 0, 1, 2$) is less severe. Z_1 gives positive indication that there is a target present. However, an ambiguity remains when the observed state is (2), whereas the true state is X_1 or X_3.

1.4.4 Numerical comparisons
The right-hand side of Table 14 gives some numerical values of:

Table 14 — A comparison between the two-target and single-target cases
$p = 0.5$, $a = 0.6931$, $(f = 0.5)$, $c = 0.5$

	One target			Two targets					
t	$P(Y_0)$	$P(Y_1)$	I_E	$P(Y_0)$	$P(Z_0)$	$P(Y_1)$	$P(Z_1)$	I_E	I_F
0.1	0.7215	0.2009	0.0866	0.9324	0.6257	0.0379	0.3375	0.0252	0.0918
0.2	0.7105	0.2133	0.0407	0.9268	0.6119	0.0422	0.3534	0.0117	0.0918
0.3	0.7028	0.2219	0.0155	0.9228	0.1023	0.0422	0.3645	0.0045	0.0376
0.4	0.6983	0.2271	0.0032	0.9203	0.5966	0.0470	0.3710	0.0009	0.0269
0.5	0.6967	0.2288	0.0007	0.9195	0.5947	0.0474	0.3731	0.0001	0.0237
0.6	0.6983	0.2271	0.0070	0.9203	0.5966	0.0467	0.3709	0.0018	0.0273
0.7	0.7028	0.2219	0.0226	0.9228	0.6023	0.0447	0.3642	0.0062	0.0383
0.8	0.7105	0.2133	0.0497	0.9268	0.6119	0.0416	0.3531	0.0143	0.0585
0.9	0.7215	0.2009	0.0948	0.9324	0.6257	0.0374	0.3373	0.0280	0.0926

Y_0 no target-like response from search, at least not apparent
Y_1 apparent target detection in correct cell

Y_0 at most one target-like response
Y_1 at least two target-like responses in correct cell(s)
Z_0 no target-like response at all
Z_1 at least one target-like response in a cell where there is a target

$$P(Y_j) = \sum_{i=1}^{3} P(X_i, Y_j)$$

$$P(Z_j) = \sum_{i=1}^{3} P(X_i, Z_j), \quad j = 0, 1,$$

and the corresponding mutual information measures

$$I_E = \sum_{i=1}^{3} \sum_{j=0}^{2} P(X_i, Y_j) \ln \left\{ \frac{P(X_i, Y_j)}{P(X_i)P(Y_j)} \right\}$$

$$I_F = \sum_{i=1}^{3} \sum_{j=0}^{2} P(X_i, Z_j) \ln \left\{ \frac{P(X_i, Z_j)}{P(X_i)P(Z_j)} \right\}$$

for the two target case with parameter $p = 0.5$, $a = 0.6931$, $c = 0.5$. For comparison a portion of Table 11 from Section 1.3 is appended on the left-hand side. It is emphasized that Y_0 and Y_1 are not the same in the two-target case as in the single-target case, so care must be taken in the comparison. It is not surprising that

$P(Y_0)$ is always greater than $P(Z_0)$, because the definitions make it that way. One sees the expected symmetries about $t = 0.5$ in the two-target case and the minimization of the information measures near $t = 0.5$. In conclusion we wish to remind the reader that this series of studies in an information-theoretic context of the effect of multiple targets, true and false, on the conduct of search, is both methodological and exploratory. Models and analysis have been developed to formulate important probabilities and information measures, both for one and two targets, against a background of unlimited false targets. Generalization to more targets is in principle possible, while further systematic investigation in connection with optimal distribution of search effort, and the link between this topic and the information generated by a search, remain to be tackled.

2

Information provided by regular surveillance of a moving target

One of the simplest and purest information-transforming processes in military operations is the surveillance of a target. Each observation of the target provides information about its location. If the target is moving, the information decays between observations and must be replenished periodically. These characteristics of the information process can be captured in some very simple models that describe the essential elements without undue detail. A quantitative description of the information dynamics can be developed from the following simple assumptions.

A target moves randomly in one dimension. At regular periodic intervals, measurements of its position are made, with variance Σ^2. The target is assumed to be detected at each observation, so there are no irregularities in the measurements.

Initially the target position is normally distributed about the origin with variance Σ^2 (resulting, for example, from the zeroth measurement). As time passes, the density flattens, but remains normal. The flattening may be described by a function of the elapsed time, t:

$$\sigma^2(t) = f(t) \, \Sigma^2 \tag{76}$$

with

$$f(0) = 1 \, ,$$
$$df/dt > 0 \; ; \text{ all } t.$$

Consequently, in the absence of measurement, the variance of the position density is always increasing.

Let the first position measurement take place at $t = \tau$. Immediately before that time, the prior density of the position x is

$$p(x) = N(x, 0, \sigma) \tag{77}$$

where

$$\sigma^2 = \sigma^2(\tau) = \Sigma^2 f(\tau), \quad N(x,\mu,\sigma) = (2\pi\sigma^2)^{-\frac{1}{2}} \exp\left(-\tfrac{1}{2}\left(\frac{x-\mu}{\sigma}\right)^2\right). \tag{78}$$

The posterior density, as a result of a measurement m, is

$$\begin{aligned} p(x|m) &= (p(m|x)\,p(x))/p(m), \\ &= (p(m|x)\,p(x))/\int p(m|x')\,p(x')\,dx'. \end{aligned} \tag{79}$$

The probability density $p(m|x)$ of obtaining a measured value m, given a target at x, is $N(m,x,\Sigma)$, and

$$\begin{aligned} p(m) &= \int_{-\infty}^{\infty} N(m,x,\Sigma)\,N(x,0,\sigma)\,dx \\ &= N(m,0,\sqrt{(\Sigma^2+\sigma^2)}). \end{aligned} \tag{80}$$

Combining (79) and (80), we get for the posterior density:

$$p(x|m) = N(m,x,\Sigma)\,N(x,0,\sigma)\,/\,N(m,0,\sqrt{(\sigma^2+\Sigma^2)}), \tag{81}$$

which turns out to be normal also, namely

$$p(x|m) = N(x, m\sigma^2/(\sigma^2+\Sigma^2), \sigma\Sigma/\sqrt{(\sigma^2+\Sigma^2)}). \tag{82}$$

This implies that the posterior density is centred at a weighted average of the prior position, 0, and the measured position, m, and has a variance $\sigma_a < \sigma_b$. (The subscripts 'b' and 'a' refer to before and after the measurements.) The measurement process thus serves to reduce the variance of the position density, in contrast to the increase due to target motion.

The probability densities can be used to derive expressions for the information. Two types of information prove to be useful. Self information of a single probability density can be used to express the decrease of information due to target motion. Mutual information between two densities can be used to express the increase in information due to the measurement process. Taken together, these two quantities define the information dynamics of the surveillance of a moving target.

The entropy of a probability density $p(x)$ is

$$H = -\int dx\, p(x)\ln p(x). \tag{83}$$

For a Gaussian density such as $N(x,0,\sigma)$ the entropy is

$$H = -\int_{-\infty}^{\infty} dx \, \frac{\exp(-x^2/2\sigma^2)}{\sigma\sqrt{2\pi}} \left[-\frac{x^2}{2\sigma^2} - \ln \sigma\sqrt{2\pi} \right] = \tfrac{1}{2} + \ln\sqrt{2\pi} + \tfrac{1}{2}\ln \sigma^2 \,. \qquad (84)$$

When σ^2 has the form specified by (78)

$$H = \tfrac{1}{2} + \ln\sqrt{2\pi} + \tfrac{1}{2}\ln \Sigma^2 + \tfrac{1}{2}\ln f = H_0 + \tfrac{1}{2}\ln f \,. \qquad (85)$$

Information I is the negative of the entropy, so, as a function of time,

$$I = I_0 - \tfrac{1}{2}\ln f(t) \,. \qquad (86)$$

This expresses the decrease in information due to target motion between measurements. The function $f(t)$ has the general characteristics implied by (76), but its specific form remains to be defined. Much of the analysis can be carried out with f as an unspecified function.

To evaluate the change in information resulting from measurement, we need to consider the probability densities both before and after the measurements, and to compute the mutual information between them. The mutual information is $\ln p(x|m)/p(x)$, and its ensemble average is

$$I = \int dx \, dm \, p(x|m) \, p(m) \, \ln(p(x|m)/p(x)) \,, \qquad (87)$$

which reduces to

$$I = \tfrac{1}{2}\ln(1 + \sigma_b^2/\Sigma^2) \,, \qquad (88)$$
$$= \tfrac{1}{2}\ln(1 + f(t)) \,, \qquad (89)$$

using the notation of (78).

Consider now a series of repeated measurements. Each supplies an amount of information

$$I_n = \tfrac{1}{2}\ln(1 + \sigma_{bn}^2/\Sigma^2) \,, \qquad (90)$$

where σ_{bn}^2 is the variance before the nth measurement. By (82), the variance after the nth measurement is

$$\sigma_{an}^2 = (\sigma_{bn}^2 \Sigma^2)/(\sigma_{bn}^2 + \Sigma^2) \,,$$

or,

$$\sigma_{an}^2/\Sigma^2 = (\sigma_{bn}^2/\Sigma^2)/(1 + \sigma_{bn}^2/\Sigma^2) \,. \qquad (91)$$

Ch. 2] **Information provided by regular surveillance of a moving target**

In addition, we assume that $f(t)$ is a universal law of density flattening in the sense that

$$\sigma^2_{bn+1} = \sigma^2_{an} f(t) , \qquad (92)$$

(all $n \geq 1$).

Combining (91) and (92), we get

$$\frac{\sigma^2_{bn+1}}{\Sigma^2} = f(\tau)(\sigma^2_{bn}/\Sigma^2)/(1 + (\sigma^2_{bn}/\Sigma^2)) , \qquad (93)$$

where τ is the time interval between measurements.

This recurrence relation can be solved explicitly in terms of f and n with the initial condition $\sigma^2_{b1}/\Sigma^2 = f(\tau)$.

Putting $f(\tau) = f$ we get

$$\sigma^2_{bn}/\Sigma^2 = (f^{n+1} - f^n)/(f^n - 1) . \qquad (94)$$

It is of interest to examine this expression in the limit $n \to \infty$. Consider first the limit $f \to 1$, corresponding to a stationary target. Putting $f = 1 + \varepsilon$, we have, to first order in ε,

$$\sigma^2_{bn}/\Sigma^2 = \frac{1 + (n+1)\varepsilon - 1 - n\varepsilon}{1 + n\varepsilon - 1} = \frac{1}{n} . \qquad (95)$$

This is the classical result,

$$\sigma^2_n \sim \Sigma^2/n \to 0 , \qquad (96)$$

for the variance of n independent, identically distributed measurements, which states that a fixed position may be specified with any desired precision, given enough measurements.

For $f > 1$, we obtain a different result:

$$\sigma^2_{bn}/\Sigma^2 = (f-1)\frac{1}{1 - \left(\frac{1}{f}\right)^n} \to f - 1 . \qquad (97)$$

Alternatively, using (92), we can write the variance after the nth measurement:

$$\sigma^2_{an}/\Sigma^2 = \left(1 - \frac{1}{f}\right) \frac{1}{1 - \left(\frac{1}{f}\right)^{n+1}} \to 1 - \frac{1}{f}. \tag{98}$$

This expresses the fact that for a moving target, even with an unlimited number of measurements, the posterior variance does not approach zero. The residual imprecision is, to first order, $\sim (f-1)\Sigma^2$, proportional to the amount of motion that occurs between measurements.

These ideas can be expressed in information terms with the use of (90). We shall be interested in three different information values: the marginal information of the nth measurement, I_n; the total information of the first N measurements,

$$I_N \equiv \sum_{1}^{N} I_n \; ;$$

and the average information per measurement, $\bar{I}_N = I_N/N$.

Equations (90) and (94) can be combined to give I_n:

$$I_n = \tfrac{1}{2} \ln \left[(f^{n+1} - 1)/(f^n - 1) \right]. \tag{99}$$

This, in turn, can be summed exactly to give

$$I_N = \tfrac{1}{2} \ln \left[(f^{N+1} - 1)/(f - 1) \right]. \tag{100}$$

In the stationary target limit ($f = 1$),

$$I_n = \tfrac{1}{2} \ln\left(1 + \frac{1}{n}\right) \tag{101}$$

$$I_N = \tfrac{1}{2} \ln(N+1) \tag{102}$$

$$\bar{I}_N = \tfrac{1}{2} \frac{\ln(N+1)}{N}. \tag{103}$$

Thus, as n and $N \to \infty$, both the marginal information and the average information per measurement approach zero. However, the total information $\to \infty$ as $\ln(N+1)$ reflecting the fact that an infinite amount of information is required to specify a position with arbitrarily great precision.

For the moving target, $f > 1$, and the limiting values are

$$I_n, \bar{I}_N \to \tfrac{1}{2} \ln f \tag{104}$$

and

$$I_N \to \frac{N}{2} \ln f \ . \tag{105}$$

In this case, the marginal and average values approach a common finite limit. By comparison with (86), that limit can be interpreted as the amount of information about the target location which is lost because of target motion between measurements, and which must be regained by each succeeding measurement. Similarly, the fact that $I_N \to \infty$ as N reflects the cumulative information losses that must be made up, even though only a finite amount of information is needed to express the target position with the precision implied by (98) at any particular time.

The sort of limiting processes discussed above imply that measurements are taken at constant intervals for an unspecified length of time. This is implicit in the assumption that f has a fixed value for given sampling interval τ.

A different problem arises in the situation for which the total length of time is fixed, and we wish to know the results of more frequent sampling within that interval. Let the total period be T, divided into N intervals. Then f is no longer a function of T, but of T/N. The total information gathered in T is

$$I_N(T) = \tfrac{1}{2} \ln[(f(T/N)^{N+1} - 1)/(f(T/N) - 1)] \ . \tag{106}$$

To evaluate this in the limit $N \to \infty$, we need some explicit form for f. On physical grounds it would be reasonable to take

$$f(T/N) = 1 + a(T/N)^\alpha \tag{107}$$

with

$$1 \leq \alpha \leq 2 \ .$$

This form for f would correspond to a random walk when $\alpha = 1$ and to a 'farthest-on' interval when $\alpha = 2$. Intermediate values of α would correspond to intermediate growth rates for the position variance. However, we do not have complete freedom to choose the form of f, because it is implicitly constrained by assumption (92), that f describes a universal law of density flattening. If (92) is applied successively we must have

$$f(2\tau) = f^2(\tau) \ . \tag{108}$$

This in turn implies that f has the form

$$f = e^{at} \ . \tag{109}$$

For small at, this is approximately $1 + at$, which describes the random walk. Using

$$f(T/N) = e^{aT/N} \tag{110}$$

in (106) we get

$$I_N(T) \sim \tfrac{1}{2} \ln N + \tfrac{1}{2} \ln \left[\frac{e^{aT} - 1}{aT} \right] . \tag{111}$$

The leading term is the same as that determined in (102) for the stationary-target limit. This leads to the following interpretation. By increasing the sampling rate indefinitely, we essentially freeze the target. The principal contribution to the information rate is equivalent to the information obtained by repeatedly measuring a fixed target.

The other term,

$$I_M(T) = \tfrac{1}{2} \ln \left[\frac{e^{aT} - 1}{aT} \right] , \tag{112}$$

represents the information about the motion obtained by continuous surveillance of a moving target for a period T.

The results of this simple model can be summarized as follows.

Target motion in the interval between regular measurements causes a decrease of information in the amount:

$$\Delta I_{TM} = -\tfrac{1}{2} \ln f . \tag{113}$$

The measurement process causes an increase in the amount of information that depends on the sequential number of the measurement:

$$\Delta I_M = +\tfrac{1}{2} \ln [(f^{n+1} - 1)/(f^n - 1)] . \tag{114}$$

For small values of the index, n, $|\Delta I_M| > |\Delta I_{TM}|$, and a net gain of information ensues. As $n \to \infty$, $|\Delta I_M| \to |\Delta I_{TM}|$, and the process comes into balance, with the information gain from the measurements equalling the information loss due to target motion. The net information gain from a long series of measurements is

$$\sum_{n=1}^{\infty} \tfrac{1}{2} \ln[(f^{n+1} - 1)/(f^n - 1)] - \tfrac{1}{2} \ln f = -\tfrac{1}{2} \ln \left(\frac{f}{f-1} \right) . \tag{115}$$

These results can form the basis for the exploration of more complex and realistic cases, such as those involving irregular measurements, missed detections, and false targets.

3

Merger of data in a filter centre

3.1 BASIC CONCEPTS

The merger of data from diverse sources is a common feature of military command and control. Filter centres are designed to collect surveillance and reconnaissance information from a variety of sensors, to merge and reconcile that information, and to forward a consolidated picture to a tactical commander. The sensors that observe the battle situation may overlap to a certain extent, but more often they are covering different aspects of a complex situation. They provide limited pieces of a puzzle that are usually complementary, occasionally contradictory, and always incomplete.

The process of merging data of this type is one of many information-modifying processes that occur within a command and control system. It is an important one, and, to understand how it interacts with the other information-modifying processes, it is necessary to have a quantitative description. This chapter develops one such quantitative description in information theoretic terms, using a relatively simple model.

The features that we wish to model can be summarized quite simply. At some instant of time $t = 0$, the relevant tactical factors of a battle situation can be totally specified by N_0 bits of information. The battle situation is observed by several sensors, $s_1, s_2, \ldots s_k$, which obtain n_j ($j = 1 \ldots k$) bits of data respectively. Usually $n_j \ll N_0$ for all k. The observations from the sensors are sent to a filter centre, where they are processed, either by a human or by some automated algorithms to create N_1 bits of information about the real battle situation.

The dominant features of the sensors are their limited coverage and their propensity for error. Thus, if the description of the real battle situation is viewed as a message N_0 bits long, then the bits reported by the sensors may be either missing (limited coverage) or incorrect (error). A convenient way to model a process with these characteristics is the use of the so-called erasure channel from communication theory.

Fig. 1 is a symbolic representation of the erasure channel. The input to the

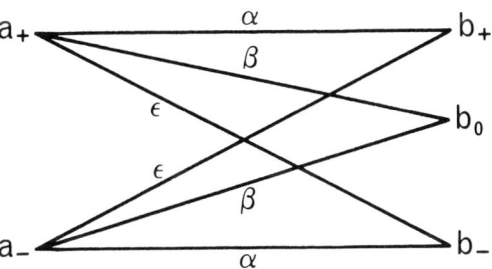

Fig. 1 — Erasure channel with errors.

channel is binary, represented as a_+ or a_-. The output is ternary, with the possibilities b_+, b_0, or b_-. The b_0 output is the 'erasure', a null result that carries no information about the input. The channel as shown is symmetric. This is not a necessary condition, but it is a realistic representation of many practical cases, and it vastly simplifies the analysis.

The quantities α, β, and ε are transition probabilities. α is the probability that an input will be correctly represented by the output; ε is the probability of a false output; and β is the probability of an inconclusive output. Naturally,

$$\alpha + \beta + \varepsilon = 1 .$$

In the analogy between the abstract channel model and a military command and control system, ε represents the probability that a sensor will return false data about the battle situation, whereas β is the probability that no data at all will be returned. $1 - \beta$ is, in fact, the effective 'coverage' provided by the sensor.

The fundamental quantity that we wish to compute on the basis of the channel model is the mutual information between the output and the input. If I_0 is the initial information content of the input (the true battle situation), I_1 is the mutual information between the output of one channel and the input, and I_2 is the mutual information between the input and the output of two parallel, independent channels, then the comparisons among I_0, I_1, and I_2 will give the necessary quantitative insights into the effects of sensor coverage (β), sensor error (ε), and fusion of data from two channels. The plan of this chapter is thus to calculate I_0, I_1, and I_2 and to compare them for various ranges of the parameters α, β, and ε.

Many probability expressions are required in the computation of the mutual information. Generally, the mutual information between two sets $\{x_j\}$, $\{y_j\}$ is:

$$I = \sum_{i,j} p(x_i, y_j) \log \frac{p(x_i, y_j)}{p(x_i) p(y_j)} . \qquad (116)$$

We identify x_i with the inputs, a_+ and a_-, and y_j with the outputs b_+, b_0, b_-.

Define

$$p(a_+) = q$$
$$p(a_-) = 1-q .\qquad(117)$$

The conditional probabilities of b given a are taken from the channel diagram

$$\alpha = p(b_+|a_+) = p(b_-|a_-)$$
$$\beta = p(b_0|a_+) = p(b_0|a_-) \qquad(118)$$
$$\varepsilon = p(b_-|a_+) = p(b_+|a_-) .$$

These are then combined to give the joint probabilities

$$\begin{aligned}p(b_+, a_+) &= \alpha q\\ p(b_+, a_-) &= \varepsilon(1-q)\\ p(b_0, a_+) &= \beta q\\ p(b_0, a_-) &= \beta(1-q)\\ p(b_-, a_+) &= \varepsilon q\\ p(b_-, a_-) &= \alpha(1-q) .\end{aligned}\qquad(119)$$

Finally, the probabilities for the b's are:

$$\begin{aligned}p(b_+) &= \alpha q + \varepsilon(1-q)\\ p(b_0) &= \beta \\ p(b_-) &= \varepsilon q + \alpha(1-q) .\end{aligned}\qquad(120)$$

All of these expressions can be combined to give the mutual information between input and output for the channel shown in Fig. 1. By (116) we obtain

$$\begin{aligned}I_1 = {}& \alpha \log\alpha + \varepsilon \log\varepsilon \\ &- (\alpha q + \varepsilon[1-q]) \log(\alpha q + \varepsilon[1-q])\\ &- (\varepsilon q + \alpha[1-q]) \log(\varepsilon q + \alpha[1-q]) .\end{aligned}\qquad(121)$$

This is a completely general expression, which reduces to various familiar special cases for particular values of the parameters.

(a) when $\alpha = 1$, $\beta = 0$, $\varepsilon = 0$, we have a perfect channel with no errors or erasures. Then

$$I_1 \to -q \log q - (1-q) \log(1-q) \equiv H(q) \qquad(122)$$

Basic concepts

which is just the source entropy. For $q = \frac{1}{2}$

$$I_1 = \log 2 = 1 \text{ bit}|\text{symbol},$$

the logarithm being to base 2.

(b) When $q = \frac{1}{2}$

$$I_1 \to \alpha \log\alpha + \varepsilon \log\varepsilon - (\alpha + \varepsilon) \log(\alpha + \varepsilon) + (\alpha + \varepsilon) \log 2 . \tag{123}$$

This further reduces to special cases:

(i) $\alpha + \varepsilon = 1$, $\beta = 0$.

This is the binary symmetric channel, for which

$$I_1 = \log 2 - H(\alpha) = 1 - H(\alpha) \tag{124}$$

(ii) $\varepsilon = 0$.

This is the traditional binary erasure channel, for which

$$I_1 = \alpha \log 2 = \alpha . \tag{125}$$

Fig. 2 shows $I_1(\alpha)$ for these two special cases. More generally it can be shown that

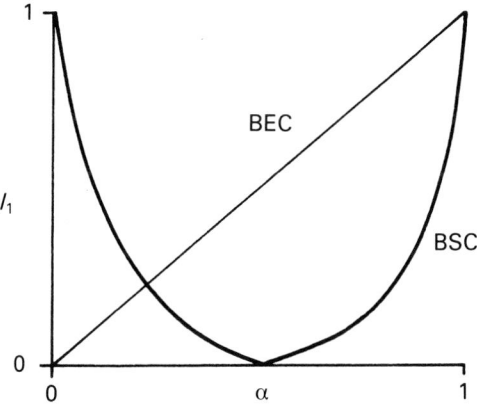

Fig. 2 — $I_1(\alpha)$ for special cases.

$I_1 \geq 0$ for all values of the parameters α, β, ε, q. As in the two examples, equality occurs for $\varepsilon = \alpha$, independently of β and q. To see this, note that

$$\frac{\partial I_1}{\partial \alpha} = \log \alpha - q \log(\alpha q + \varepsilon[1-q]) - (1-q) \log(\varepsilon q + \alpha[1-q]) \qquad (126)$$

and

$$\frac{\partial I_1}{\partial \varepsilon} = \log \varepsilon - (1-q) \log(\alpha q + \varepsilon[1-q]) - q \log(\varepsilon q + \alpha[1-q]) . \qquad (127)$$

When we set $\partial I_1/\partial \alpha = \partial I_1/\partial \varepsilon = 0$ we get

$$q(1-q)(\alpha - \varepsilon)^2 = 0 \qquad (128)$$

as the condition for stationarity. It is then easily verified that $I_1 = 0$ when $\alpha = \varepsilon$. Table 15 shows representative values of I_1 for various values of α, ε for $q = 0.4$.

Table 15 — Representative values of I_1
$q = 0.4$

ε \ α	0	0.1	0.2	0.3	0.4	0.5	0.6	0.7	0.8	0.9	1.0
0.0	0	0.097	0.194	0.291	0.388	0.485	0.583	0.680	0.777	0.874	0.971
0.1	0.097	0	0.023	0.073	0.134	0.202	0.276	0.352	0.431	0.513	
0.2	0.194	0.023	0	0.139	0.047	0.092	0.145	0.204	0.268		
0.3	0.291	0.073	0.014	0	0.010	0.035	0.071	0.114			
0.4	0.388	0.134	0.047	0.010	0	0.008	0.028				
0.5	0.485	0.202	0.092	0.035	0.008	0					
0.6	0.583	0.276	0.145	0.071	0.028						
0.7	0.680	0.352	0.204	0.114							
0.8	0.777	0.431	0.268								
0.9	0.874	0.513									
1.0	0.971										

The calculation of I_2 proceeds in the same manner, although the details become more complicated. We assume that two channels each provide outputs. If those are considered pairwise, the relevant probabilities of the possible outcomes are:

$$\begin{aligned} p(b_{++}|a_+) &= p(b_{--}|a_-) = \alpha_1 \alpha_2 \\ p(b_{+0}|a_+) &= p(b_{-0}|a_-) = \alpha_1 \beta_2 + \alpha_2 \beta_1 \\ p(b_{+-}|a_+) &= p(b_{+-}|a_-) = \alpha_1 \varepsilon_2 + \alpha_2 \varepsilon_1 \\ p(b_{00}|a_+) &= p(b_{00}|a_-) = \beta_1 \beta_2 \end{aligned} \qquad (129)$$

Sec. 3.1] **Basic concepts** 61

$$p(b_{0-}|a_+) = p(b_{+0}|a_-) = \beta_1 \varepsilon_2 + \beta_2 \varepsilon_1$$
$$p(b_{--}|a_+) = p(b_{++}|a_-) = \varepsilon_1 \varepsilon_2 .$$

The joint probabilities of the a's and b's and the unconditional probabilities of the b's can then be formed. When they are all incorporated properly in (116), the result is:

$$I_2 = \alpha_1 \alpha_2 \log \alpha_1 \alpha_2 + \varepsilon_1 \varepsilon_2 \log \varepsilon_1 \varepsilon_2$$
$$- B \log B - C \log C + D \log D + E \log E - F \log F - G \log G \quad , \quad (130)$$

where

$$B = \alpha_1 \alpha_2 q + \varepsilon_1 \varepsilon_2 (1-q)$$
$$C = \alpha_1 \alpha_2 (1-q) + \varepsilon_1 \varepsilon_2 q$$
$$D = \alpha_1 \beta_2 + \alpha_2 \beta_1$$
$$E = \beta_1 \varepsilon_2 + \beta_2 \varepsilon_1$$
$$F = (\alpha_1 \beta_2 + \alpha_2 \beta_1) q + (\beta_1 \varepsilon_2 + \beta_2 \varepsilon_1)(1-q)$$
$$G = (\beta_1 \varepsilon_2 + \beta_2 \varepsilon_1) q + (\alpha_1 \beta_2 + \alpha_2 \beta_1)(1-q) .$$

This is a form that allows numerical calculations for specific cases to be made. However, with seven parameters, it is too complicated to yield any general results, so we will examine more restricted cases.

First, when the two channels are identical, I_2 reduces to

$$I_2 = 2\beta I_1 + \alpha^2 \log \alpha^2 + \varepsilon^2 \log \varepsilon^2$$
$$- (\alpha^2 q + \varepsilon^2(1-q)) \log(\alpha^2 q + \varepsilon^2(1-q))$$
$$- (\varepsilon^2 q + \alpha^2(1-q)) \log(\varepsilon^2 q + \alpha^2(1-q)) . \quad (131)$$

For general comparison with I_1, we can then look at the same special cases as were considered above.

        (a) $\alpha = 1$, $\beta = \varepsilon = 0$.   The perfect channel.

Again

$$I_2 \to H(q)$$

        (b) $q = \frac{1}{2}$

            (i) $\beta = 0$

$$I_2 \to (\alpha^2 + (1-\alpha)^2) \log 2 + \alpha^2 \log \alpha^2 + (1-\alpha)^2 \log(1-\alpha)^2$$
$$- (\alpha^2 + (1-\alpha)^2) \log(\alpha^2 + (1-\alpha)^2) \ . \tag{132}$$

This has a shape similar to that of I_1, but differs slightly. Fig. 3 plots both $I_1(\alpha)$ and

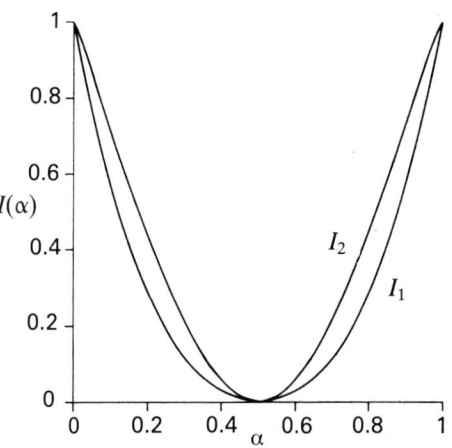

Fig. 3 — Plot of I_1 and I_2.

$I_2(\alpha)$ for this special case.

(ii) $\varepsilon = 0$

In this case $I_2 \to 2\alpha - \alpha^2$ which is compared with I_1 in Fig. 4.

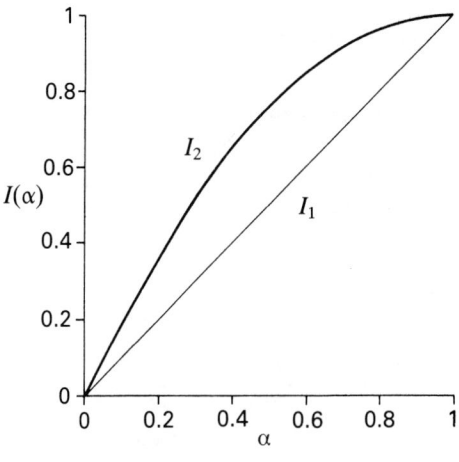

Fig. 4 — Plot of I_1 and I_2.

Sec. 3.1] Basic concepts 63

Figs. 3 and 4 both show that $I_2 \geq I_1$ for these special cases. These are suggestive of the general result.

Direct calculation of I_2 is shown in Table 16 for a range of values of α and ε. Table

Table 16 —Representative values of I_2
$q = 0.4$

ε \ α	0	0.1	0.2	0.3	0.4	0.5	0.6	0.7	0.8	0.9	1.0
0.0	0	0.184	0.350	0.495	0.621	0.728	0.816	0.883	0.932	0.961	0.971
0.1	0.184	0	0.046	0.138	0.245	0.354	0.459	0.557	0.644	0.719	
0.2	0.350	0.046	0	0.028	0.091	0.173	0.263	0.355	0.445		
0.3	0.495	0.138	0.028	0	0.020	0.068	0.135	0.211			
0.4	0.621	0.245	0.091	0.020	0	0.015	0.056				
0.5	0.728	0.354	0.173	0.068	0.015	0					
0.6	0.816	0.459	0.263	0.135	0.056						
0.7	0.883	0.557	0.355	0.211							
0.8	0.932	0.644	0.445								
0.9	0.961	0.719									
1.0	0.971										

17 shows $\Delta = I_2 - I_1$. From these, we can see that $\Delta \geq 0$ for all parameter values, with

Table 17 — Representative values of $\Delta = I_2 - I_1$
$q = 0.4$

ε \ α	0	0.1	0.2	0.3	0.4	0.5	0.6	0.7	0.8	0.9	1.0
0.0	0	0.087	0.156	0.204	0.233	0.243	0.233	0.203	0.155	0.087	0
0.1	0.087	0	0.023	0.066	0.111	0.152	0.184	0.205	0.213	0.207	
0.2	0.156	0.023	0	0.014	0.044	0.081	0.118	0.151	0.177		
0.3	0.204	0.066	0.014	0	0.010	0.033	0.064	0.097			
0.4	0.233	0.111	0.044	0.010	0	0.008	0.027				
0.5	0.243	0.152	0.081	0.033	0.008	0					
0.6	0.233	0.184	0.118	0.0614	0.027						
0.7	0.203	0.205	0.151	0.097							
0.8	0.155	0.213	0.177								
0.9	0.087	0.207									
1.0	0										

equality occurring only at boundaries and where $\alpha = \varepsilon$. Consequently, by the criterion of information, merger of two data streams is always advantageous, regardless of the size of the errors or the number of erasures. When other criteria are used, the situation is not so clearcut, however.

In the practical situation of merging tactical data and trying to make tactical decisions, it is important to know whether better decisions can be made on the basis of the merged or the unmerged data. Thus, the probability of correctly interpreting a received symbol is important, as is the probability of making an error. Since we allow the possibility of erasures and, in merged data, of contradictions, the probabilities of being right and of being wrong do not add to 1.

The calculation of these probabilities presupposes some decoding algorithm which assigns a unique value of the a symbol to each value of the observed b symbol or combinations thereof. We choose for illustration a very simple and highly practical assignment algorithm.

For one channel we use:

Observed	Assigned
+	+
0	None
−	−

For two channels we use:

Observed	Assigned
+ +	+
+ 0	+
+ −	None
0 +	+
0 0	None
0 −	−
− +	None
− 0	−
− −	−

For higher numbers of channels we follow the same basic rules, of assigning a + when the number of observed plusses is greater than the number of observed minuses, assigning a − when the number of assigned minuses is greater, and making no assignment when + 's and − 's are equal or all zero. When these rules are combined with the probabilities of (118) and (129) and with the analogue for three channels, we get:

No. of channels	Probability of correct decoding P_c	Probability of error P_e
1	α	ε
2	$\alpha(\alpha + 2\beta)$	$\varepsilon(\varepsilon + 2\beta)$
3	$\alpha(\alpha^2 + 3\alpha\varepsilon + 3\alpha\beta + 3\beta^2)$	$\varepsilon(\varepsilon^2 + 3\varepsilon(1-\varepsilon) + 3\beta^2)$

If merger of data is uniformly superior to the use of unmerged data, we would expect

$$P_{3c} > P_{2c} > P_{1c}$$

and

$$P_{3e} < P_{2e} < P_{1e}.$$

That is, as data from more channels are added, we would ideally want the probability of being correct to increase and the probability of error to decrease. To determine the conditions under which this is the case, we examine the relationships among the pairwise probabilities in the α, β, ε space. For this purpose, it is most convenient to eliminate α and to work in the $\beta - \varepsilon$ plane. In that plane, we get, by equating the various probabilities:

For $P_{1c} = P_{2c}$:

$$\alpha + 2\beta = 1$$
or
$$\beta = \varepsilon.$$
(133)

For $P_{2c} = P_{3c}$:

$$\alpha + 2\beta = \alpha^2 + 3\alpha(\varepsilon + \beta) + 3\beta^2$$
or
$$\varepsilon = \tfrac{1}{2} - \beta + \sqrt{(\tfrac{3}{2}\beta^2 - \beta + \tfrac{1}{4})}.$$
(134)

For $P_{1c} = P_{3c}$:

$$\alpha^2 + 3\alpha(\varepsilon + \beta) + 3\beta^2 = 1$$
or
$$\varepsilon = \tfrac{1}{4} - \beta + \tfrac{1}{4}\sqrt{(1 + 24\beta^2)}.$$
(135)

These relations are plotted in Fig. 5 in the region $0 \leq \beta \leq 1$, $0 \leq \varepsilon \leq 1$. The figure shows that the boundary lines divide the area into six segments, corresponding to the six possible orderings of P_{1c}, P_{2c}, and P_{3c}. In only one of these segments, for $\beta > \varepsilon$ and for $\varepsilon \leq 0.33$ is the desired ordering achieved.

The relations among the P_e values can be used in a similar manner.

For $P_{1e} = P_{2e}$:

$$\varepsilon = 1 - 2\beta.$$
(136)

For $P_{2e} = P_{3e}$:

$$\varepsilon + 2\beta = \varepsilon^2 + 3\varepsilon(1 - \varepsilon) + 3\beta^2$$
or
(137)

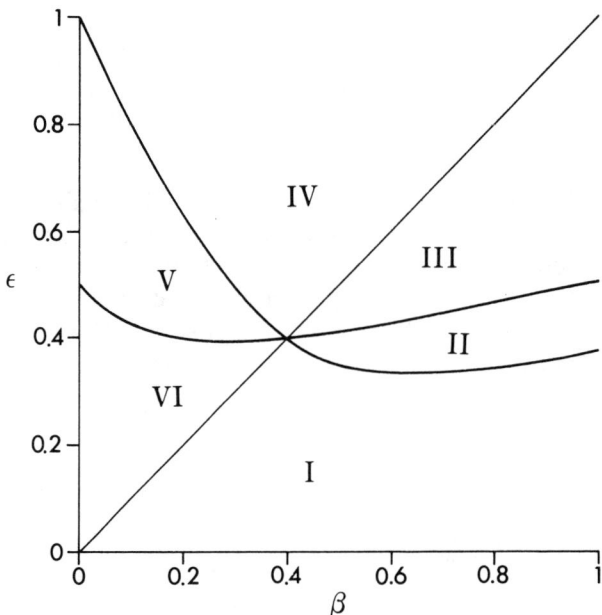

Fig. 5 — Relations between P_{1c} P_{2c}, P_{3c}
I: $P_{3c} > P_{2c} > P_{1c}$
II: $P_{2c} > P_{3c} > P_{1c}$
III: $P_{2c} > P_{1c} > P_{3c}$
IV: $P_{1c} > P_{2c} > P_{3c}$
V: $P_{1c} > P_{3c} > P_{2c}$
VI: $P_{3c} > P_{1c} > P_{2c}$

$$\varepsilon = \tfrac{1}{2} \pm \tfrac{1}{2} \sqrt{(6\beta^2 - 4\beta + 1)}.$$

For $P_{1e} = P_{3e}$:

$$1 = \varepsilon^2 + 3\varepsilon(1-\varepsilon) + 3\beta^2$$

or (138)

$$\varepsilon = \tfrac{3}{4} - \tfrac{1}{4}\sqrt{(1 + 24\beta^2)}.$$

These curves are used to divide the $\beta - \varepsilon$ plane as shown in Fig. 6. Again, only one relatively small region has the desired ordering.

Fig. 7 shows the intersection of the areas for which both desired orderings are found for the case of two channels:

$$P_{1c} < P_{2c}$$
and
$$P_{2e} < P_{1e}.$$

It is a small part of the allowed space for β and ε, and we can conjecture that the

Sec. 3.2]   **Generalized theory**   67

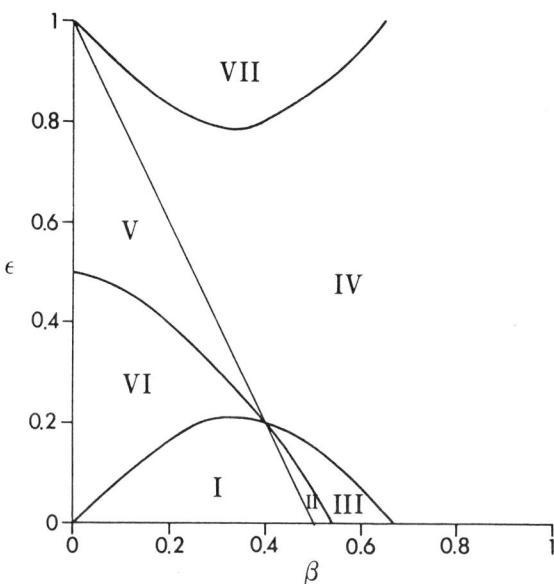

Fig. 6 — Relations between P_{1e}, P_{2e}, P_{3e}.
I: $P_{3e} < P_{2e} < P_{1e}$
II: $P_{3e} < P_{1e} < P_{2e}$
III: $P_{1e} < P_{3e} < P_{2e}$
IV: $P_{1e} < P_{2e} < P_{3e}$
V: $P_{2e} < P_{1e} < P_{3e}$
VI: $P_{2e} < P_{3e} < P_{1e}$
VII: $P_{1e} < P_{3e} < P_{2e}$

corresponding segment for a larger number of channels would be even smaller. Consequently, if we try to impose a strict criterion based on the ordering of probabilities of being right and wrong, we find that the merging of data is rarely universally superior to the use of unmerged data from a single source. The use of such a criterion should not be imposed blindly, however. In practical applications, one is always concerned with the degree of change in the probabilities achievable through merging of data, not necessarily with strict ordering. Tables 18 to 20 give some examples.

Table 18 shows a case for which the probability of being right increases significantly as additional channels are merged. The probability of being wrong also increases, but both the absolute magnitude and the amount of change are small. In this case, a prudent decision-maker would choose to use the added data.

Table 19 shows the ideal case in which the probability of being right increases and the probability of being wrong decreases as more channels are merged.

Table 20 is the problematic case. The probability of being right decreases as the second channel is added and then increases above the initial value when the third channel is added. Conversely, the probability of being wrong decreases as the second channel is added but increases again with the third channel. Moreover, the absolute magnitude of P_e is quite significant in all cases.

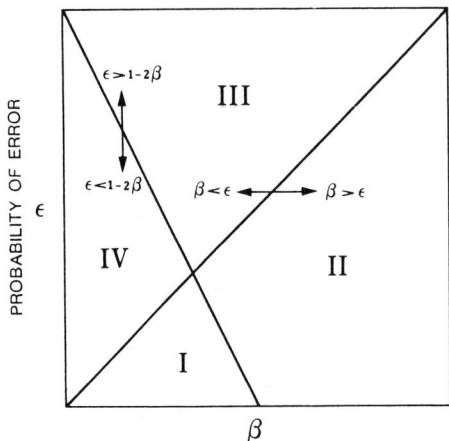

Fig. 7 — Joint ordering of $P_{1c}, P_{2c}, P_{1e}, P_{2e}$ (two channels).

I: $P_{2c} > P_{1c}$ and $P_{2e} < P_{1e}$

II: $P_{2c} > P_{1c}$ but $P_{2e} > P_{1e}$

III: $P_{2c} < P_{1c}$ and $P_{2e} > P_{1e}$

IV: $P_{2c} < P_{1c}$ but $P_{2e} < P_{1e}$

Table 18
Case: $\beta = 0.8$; $\alpha = 0.18$; $\varepsilon = 0.02$

Channels	P_c	P_e	$P_?$
1	0.18	0.02	0.80
2	0.32	0.03	0.65
3	0.43	0.04	0.53

Table 19
Case: $\beta = 0.4$; $\alpha = 0.54$; $\varepsilon = 0.06$

Channels	P_c	P_e	$P_?$
1	0.54	0.06	0.40
2	0.72	0.05	0.23
3	0.82	0.04	0.14

Table 20
Case: $\beta = 0.20$; $\alpha = 0.50$; $\varepsilon = 0.30$

Channels	P_c	P_e	$P_?$
1	0.50	0.30	0.20
2	0.45	0.21	0.34
3	0.56	0.25	0.19

3.2 GENERALIZED THEORY

In this section we generalize the model developed in Section 3.1 and provide an analysis which will facilitate further investigation. The conclusions reported above hold good concerning the sometimes curious response to added information of the probabilities of correct, incorrect, and null inferences regarding the message. Some tables are given which include results for more observers with their corresponding independent channels.

In addition we investigate the effect of a single observer who chooses to transmit his interpretation of a message down a number of independent channels in the belief that this should somehow enhance the credibility of what arrives.

It is convenient to build more detail into the model. This clarifies the structure of the probabilities α, β, ε, introduced above, according to a particular interpretation of the processes of observation and transmission.

3.2.1 Case 1: n independent observers each with an independent communication channel to the fusion centre

In the following we introduce notation different from that used in Section 3.1. This is to identify the details of the structure. It is supposed that an observer in the field perceives a message X which has either the value $X_0 = 0$, or $X_1 = 1$. With probability c the observer decodes, or identifies, the message correctly: with probability $\bar{c} = 1 - c$ the message is identified as the complement of what it actually is, namely X_0 is decoded as X_1, and conversely. The decoded message is then passed down an information channel to a fusion centre whose task is to collect and reconcile information received from diverse sources and to pass an interpreted version to a decision-maker. The information channel may be serviceable with probability s, or unserviceable with probability $\bar{s} = 1 - s$. A channel may be unserviceable because the equipment is out of order, because a part of the route on which it depends is out of action or destroyed, or because of jamming; a host of other reasons can be imagined according to the context. If serviceable, a channel may be defective in another fashion; it may 'garble' the information supplied to it in such a way that the recipient receives a false impression. Degrees of garbling and identification belong to the fashionable and improbable realm of so-called fuzzy logic. Here we merely say that with probability g the sense of the message imparted by the observer at the input is reversed, so that if X_0 were transmitted X_1 would be received, and vice versa. With probability $\bar{g}$ the message transmitted is received, and $\bar{g} = 1 - g$. It is thus seen that the probabilities α, β, ε introduced in Section 3.1 for a single channel are given by

$$\alpha = sr, \qquad \varepsilon = s\bar{r}, \qquad \beta = \bar{s}, \qquad (139)$$

where

$$r = c\bar{g} + \bar{c}g, \qquad \bar{r} = cg + \bar{c}\bar{g},$$

and r is the probability that the correct message is received, whether or not it was correctly interpreted by the observer, conditional on the channel being in service. $\bar{r} = 1 - r$ is the probability that the wrong message is received.

Given that there are n independent observers and corresponding channels all possessing the foregoing characteristics the number of correct messages received at the fusion centre is a random variable K whose probability $p_k = P(K=k)$ is the coefficient of x^k in the generating function

$$E(x^K) = \bar{s}^n + \binom{n}{1}\bar{s}^{n-1}s(rx+\bar{r}) + \binom{n}{2}\bar{s}^{n-2}s^2(rx+\bar{r}) + \ldots + \binom{n}{n}s^n(rx+\bar{r})^n$$
$$= (\bar{s}+s(rx+\bar{r}))^n = (1-sr+srx)^n. \qquad (140)$$

The independence of the channels guarantees the correctness of this result. It is not, however, quite what we want for the purpose of this analysis.

If at least one message is received (that is, if at least one channel is serviceable) an inference can be made at the fusion centre concerning the nature of the message X. The algorithm prescribed in Section 3.1 is to accept the majority, that is to accept the interpretation provided by the greater number of channels. The third possibility, referred to in Section 3.1 as 'erasure', is the acknowledgement of impossibility to make an inference; this occurs if either all channels are unserviceable, so that no transmission at all is received, or that an even number is serviceable and impart an equal number of either interpretation of the message. We call the probability of this event M_n, and it is easily seen from (140), or otherwise, to be given by

$$M_n = \bar{s}^n + \binom{n}{2}\bar{s}^{n-2}s^2\binom{2}{1}(r\bar{r}) + \binom{n}{4}\bar{s}^{n-4}s^4\binom{4}{2}(r\bar{r})^2 + \ldots +$$
$$+ \binom{n}{n_1}\bar{s}^{n-n_1}s^{n_1}\binom{n_1}{n_1/2}(r\bar{r})^{n_1/2} \qquad (141)$$

where n_1 is the largest even integer which does not exceed n. Let L_n be the probability that the incorrect inference is drawn. This is the sum of terms obtained by considering in turn the possibility that $m(=1, 2, \ldots, n)$ channels are serviceable. Thus

Generalized theory

$$L_n = \sum_{m=1}^{n} \binom{n}{m} \bar{s}^{n-m} s^m (r^m + \binom{m}{1} \bar{r}^{m-1} r + \binom{m}{2} \bar{r}^{m-2} r^2 + \ldots +$$
$$\binom{m}{m_2} \bar{r}^{m-m_2} r^{m_2}) \tag{142}$$

where $m_2 = (m/2) - 1$ *if* m is even and $(m-1)/2$ if m is odd. There is a corresponding symmetrical expression for N_n, the probability that the correct inference is drawn, in which r and $\bar{r}$ are interchanged. Clearly $L_n + M_n + N_n = 1$.

Let us now introduce the generating function $g_n(x)$ defined by

$$g_n(x) = (s\bar{r}/x + \bar{s} + srx)^n = \sum_{-n}^{n} g_m^{(n)} x^m, \tag{143}$$

say. Then it is easily checked that

$$L_n = \sum_{-n}^{-1} g_m^{(n)}, \quad M_n = g_0^{(n)}, \quad N_n = \sum_{1}^{n} g_m^{(n)}. \tag{144}$$

The conditional probabilities PR, PW of making a correct or incorrect inference are then

$$PW = L_n/(1-M_n), \quad PR = N_n/(1-M_n). \tag{145}$$

Finally we need the mean increase in information induced by any transmissions received. This is based on the joint probabilities $P(X_i, Z_j)$, where X_i is the message and Z_j is the inference. We define the inferences as Z_0 that the message is X_0, Z_1 that it is X_1, and Z_2 is reserved for 'don't know'.

Let

$$P(X = X_1) = p, \quad P(X = X_0) = q = 1-p. \tag{146}$$

Then the mean increase in information induced by the transmissions is given by

$$I_e = N_n \log N_n + L_n \log L_n - (qN_n + pL_n)\log(qN_n + pL_n) - \\ - (qL_n + pN_n)\log(qL_n + pN_n). \tag{147}$$

Given the form (143) which generates the probabilities of interest it is particularly easy to evaluate them numerically. Unfortunately the same cannot be said of the algebra. Put

$$f_n(\theta) = g_n(e^{i\theta}) . \qquad (148)$$

Then $f_n(\theta)$ is a finite Fourier series and therefore

$$g_m^{(n)} = \frac{1}{2\pi} \int_0^{2\pi} e^{-im\theta} f_n(\theta) \, d\theta , \qquad (-n \leq m \leq n) . \qquad (149)$$

Since the series is finite it follows that if we choose any positive integer t greater than n, the integral in (149) is exactly equal to a discrete Fourier transform type of series, that is

$$g_m^{(n)} = \frac{1}{t} \sum_{k=0}^{t-1} e^{-2\pi i m k/t} f(2\pi k/t) \qquad (-n \leq m \leq n) . \qquad (150)$$

This means that fast Fourier transform algorithms provide a quick way of evaluating the fundamental probabilities.

Finally, of mathematical interest is the fact that $g_m^{(n)}$ can be expressed in terms of associated Legendre functions, at least formally.

Hobson's definition of the function concerned is given in [12], Chapter XV. Thus

$$P_n^m(z) = (z^2 - 1)^{m/2} \, d^m P_n(z)/dz^m$$

where m is a positive integer and n is unrestricted, while $P_n(z)$, the coefficient of h^n in $(1 - 2hz + h^2)^{-\frac{1}{2}}$, is the Legendre Polynomial given by

$$P_n(z) = \sum_{r=0}^{m} \frac{(-)^r (2n - 2r)!}{2^n r!(n - r)!(n - 2r)!} z^{n-2r} ,$$

where $m = n/2$ or $(n - 1)/2$, whichever is an integer. [12] shows that $P_n^m(z)$ has the integral representation

$$P_n^m(z) = \frac{(n+1)(n+2) \ldots (n+m)}{2\pi} \int_0^{2\pi} (z + (z^2 - 1)^{\frac{1}{2}} \cos\phi)^n \, e^{-im\phi} \, d\phi ,$$

for all z.

After some transformation of (149) we obtain the formula

Sec. 3.2] Generalized theory

$$g_m^{(n)} = \frac{R^n}{(n+1)(n+2)\ldots(n+m)} (r/\bar{r})^{m/2} P_n^m(\bar{s}/R)$$

$$(m = 0, 1, \ldots n)$$

where

$$R^2 = \bar{s}^2 - 4r\bar{r}\bar{s}^2 . \tag{151}$$

$g_m^{(n)}$ for negative m is obtained by merely interchanging r and $\bar{r}$.

Since the asymptotic properties of $P_n^m(z)$ are available these are of potential help in providing information about $g_m^{(n)}$.

Equation (149) provides a way of investigating the effect on erasure probability of adding an extra channel. This can be done by examining the sign of $M_{n+1} - M_n$, the difference between erasure probabilities, or loss. Recalling that $M_n = g_0^{(n)}$, we have

$$M_{n+1} - M_n = g_0^{(n+1)} - g_0^{(n)} = \frac{1}{2\pi} \int_0^{2\pi} f_n(\theta) \, (s\bar{r}e^{-i\theta} + \bar{s} + sre^{i\theta} - 1) \, d\theta$$

$$= s(\bar{r} g_1^{(n)} + r g_{-1}^{(n)} - M_n) .$$

Because of symmetry this can be written

$$M_{n+1} - M_n = s(2\bar{r} g_1^{(n)} - M_n) . \tag{152}$$

Next we record $g_1^{(n)}$ for even and odd values of n, $g_0^{(n)}$ having been given by (141)

$n = 2N$:

$$g_1^{(2N)} = \binom{2N}{N}\binom{N}{1} s^{2N-1} \bar{s} \, \bar{r}^{N-1} r^N +$$

$$\binom{2N}{N+1}\binom{N+1}{3} s^{2N-3} \bar{s}^3 \bar{r}^{N-2} r^{N-1} +$$

$$+ \binom{2N}{N+2}\binom{N+2}{5} s^{2N-5} \bar{s}^5 \bar{r}^{N-3} r^{N-2} + \ldots + \binom{2N}{2N-1} s \, \bar{s}^{2N-1} r .$$

$$\tag{153}$$

$n = 2N+1$:

$$g_1^{(2N+1)} = \binom{2N+1}{N} s^{2N+1} \bar{r}^N r^{N+1} +$$

$$\binom{2N+1}{N+1}\binom{N+1}{2} s^{2N-1} \bar{s}^2 \bar{r}^{N-1} r^N +$$

$$+ \binom{2N+1}{N+2}\binom{N+2}{4} s^{2N+3} \bar{s}^4 r^{N-2} + \ldots + \binom{2N+1}{2N} s \bar{s}^{2N} r \ . \quad (154)$$

Suppose that the channels are highly reliable. This means that $s \to 1$ and the dominant terms in (153) and (154) are those corresponding to the higher powers of s.

Writing $\underset{D}{=}$ to mean 'dominated by', we find that

$$M_{2N+1} - M_{2N} \underset{D}{=} \binom{2N}{N} (\bar{r}\bar{r}s^2)^N (2N\bar{s} - s) \ ,$$

and

$$M_{2N+2} - M_{2N+1} \underset{D}{=} \binom{2N+1}{N} s (\bar{r}\bar{r}s^2)^N (2r\bar{r}\bar{s} - (N+1)\bar{s}) \ . \quad (155)$$

If adding an extra observer results in a reduction in the erasure probability it could be said that the expenditure was worthwhile. However, it is seen for high s (~ 0.9) and moderate $r(\sim 0.5)$, that the addition of an observer to an already odd number eventually makes even worse than odd, and adding a channel to an even number results in an odd number superior to even. An example is given below with $s = 0.9$, $r = \bar{r} = 0.5$.

Then $s/\bar{s} = 9$, $2r\bar{r}s/\bar{s} = 4.5$.

n			
1			
2	$M_2 > M_1$	since	$1 < 4.5$: Bad
3	$M_3 < M_2$		$2 < 9$: Good
4	$M_4 > M_3$		$2 < 4.5$: Bad
5	$M_5 < M_4$		$4 < 9$: Good
6	$M_6 > M_5$		$3 < 4.5$: Bad
7	$M_7 < M_6$		$6 < 9$: Good
8	$M_8 > M_7$		$4 < 4.5$: Good
9	$M_9 < M_8$		$8 < 9$: Good
10	$M_{10} < M_9$		$5 > 4.5$: Good
11	$M_{11} > M_{10}$		$10 > 9$: Bad
12	$M_{12} < M_{11}$		$6 > 4.5$: Good
13	$M_{13} > M_{12}$		$12 > 9$: Bad!

Generalized theory

A similar procedure can be used to make rough statements about the effect on N_n of using the messages of additional observers. We get

$$N_{n+1} - N_n = s(r\, M_n - \bar{r}\, g_1^{(n)}) \, .$$

Adding an extra channel is by this criterion 'better' if $N_{n+1} > N_n$ since N_n is the probability of making a correct inference at the fusion centre. The values of s and r used above give results consistent with those noted above.

An explanation for this apparently bizarre behaviour can be furnished if we assume that the network is almost fault-free ($s \doteq 1$), though garbling may still occur as well as observer unreliability (c). Now

$$M_{2N} \doteq s^{2N} \binom{2N}{N} (r\bar{r})^N \, ,$$

$$M_{2N+1} \doteq (2N+1)\bar{s}s^{2N}\binom{2N}{N}(r\bar{r})^N \, ,$$

and putting $s = 1$ we have $M_n \neq 0$ for even channels, and yet zero for an odd number. The reason is clear. There can only be an erasure when the number of channels is even. It is therefore not surprising that when s is very close to unity $M_{n+1} - M_n$ should alternate in sign: nor, moreover, that for a highly reliable system an odd number of channels is preferable to an even number in yielding an extremely low probability of erasure.

For large values of N and s close to 1, we also have, by use of Stirling's formula,

$$M_{2N} \sim (4r\bar{r}s^2)^N/(\pi N)^{\frac{1}{2}} \, , \qquad M_{2N+1} \sim (2N+1)\,\bar{s}\, M_{2N} \, . \tag{156}$$

Similarly, when $s = 1$,

$$N_{2N} = r^{2N} + \binom{2N}{1} r^{2N-1}\bar{r} + \ldots + \binom{2N}{N-1} r^{N+1}\bar{r}^{N-1} \, .$$

$$N_{2N+1} = r^{2N+1} + \binom{2N+1}{1} r^{2N}\bar{r} + \ldots + \binom{2N+1}{N} r^{N+1}\bar{r}^N \, .$$

The sums can be approximated for large N by using the normal approximation to the binomial. For even $n = 2N$ we have, as $N \to \infty$,

$$P\left(\alpha \leq \frac{m - 2Nr}{\sqrt{(2Nr\bar{r})}} \leq \beta\right) \sim \frac{1}{\sqrt{(2\pi)}} \int_\alpha^\beta e^{-\frac{1}{2}x^2}\, dx \, .$$

Now

$$N_{2N} = P(N+1 \leq m \leq 2N) \sim \frac{1}{\sqrt{(2\pi)}} \int_\alpha^\beta e^{-\frac{1}{2}x^2} dx$$

where

$$\beta = \left(\frac{2N\bar{r}}{r}\right)^{1/2}, \quad \alpha = \frac{N(\bar{r}-r)+1}{(2N r\bar{r})^{1/2}}.$$

Then, if

$$r > \tfrac{1}{2}, \quad N_{2N} \to 1$$
$$r = \tfrac{1}{2}, \quad N_{2N} \to \tfrac{1}{2}$$
$$r < \tfrac{1}{2}, \quad N_{2N} \to 0$$

and for s very close to unity a factor s^{2N} can be inserted on the right-hand sides for evaluation for particular finite but large N.

Similar results are obtained for odd n, and let it not be forgotten that the model provides for a very bad observer who, with an equally bad case of noise on the channel, can obtain as good results as an excellent observer with crystal transmission clarity.

Conversely, when the channels in the network have a large unserviceability probability ($\bar{s} \to 1$), M_n is dominated by $\bar{s}^n$, monotonic decreasing with n. On the other hand,

$$N_n \underset{D}{=} \bar{s}^{n-1} s r n,$$

and the probability of a correct inference at the fusion centre, conditional on there having been no erasure, is for even $n = 2N$,

$$PR = \frac{N_{2N}}{1 - M_{2N}} = \frac{2N\bar{s}^{2N-1} s r}{\left\{1 - s^{2N}\binom{2N}{N}(r\bar{r})^N\right\}},$$

and for odd $n = 2N+1$,

Sec. 3.2]   **Generalized theory**   77

$$PR = \frac{(2N+1)\bar{s}^{2N} s r}{\left\{1 - (2N+1)\bar{s}\, s^{2N} \binom{2N}{N} (\overline{rr})^N\right\}}.$$

In both cases it is the numerator which dominates for $\bar{s} \doteq 1$.

For non-extreme s and $n \to \infty$ we can conveniently use an exact representation for Legendre functions given by Abramowitz & Stegun [13]. The formula is

$$P_\nu^\mu(z) = (2\pi)^{-\frac{1}{2}} (z^2-1)^{-\frac{1}{4}} \frac{\Gamma(\nu+\mu+1)}{\Gamma(\nu+\frac{3}{2})} \{[z+(z^2-1)^{\frac{1}{2}}]^{\nu+\frac{1}{2}} F(\tfrac{1}{2}+\mu, \tfrac{1}{2}-\mu;$$

$$\tfrac{3}{2}+\nu; Z_1) + ie^{-i\mu\pi}[z-(z^2-1)^{\frac{1}{2}}]^{\nu+\frac{1}{2}} F(\tfrac{1}{2}+\mu, \tfrac{1}{2}-\mu; \tfrac{3}{2}+\nu; -Z_2)\},$$

where

$$Z_1 = \frac{z+(z^2-1)^{\frac{1}{2}}}{2(z^2-1)^{\frac{1}{2}}}, \qquad Z_2 = \frac{z-(z^2-1)^{\frac{1}{2}}}{2(z^2-1)^{\frac{1}{2}}}. \tag{157}$$

This enables any of the coefficients $g_m^{(n)}$ to be expressed in this way. In particular, the leading terms in M_n are obtained from (157) with $\mu = 0$. Then

$$M_n = R^n P_n(\bar{s}/R) = \frac{1}{\{2\pi(4\overline{rr}s^2)^{\frac{1}{2}}\}^{\frac{1}{2}}} \frac{\Gamma(n+1)}{\Gamma(n+\frac{3}{2})} \{\bar{s}+2s(\overline{rr})^{\frac{1}{2}}\}\left\{1 + \frac{\frac{1}{4}Z_1}{(\frac{3}{2}+n)} + \ldots\right\}$$

where

$$Z_1 = \tfrac{1}{2}\frac{\bar{s}+2s(\overline{rr})^{\frac{1}{2}}}{2s(\overline{rr})^{\frac{1}{2}}} = \tfrac{1}{2}\left\{1+\frac{\bar{s}}{2s(\overline{rr})^{\frac{1}{2}}}\right\}, \tag{158}$$

and R is given by (151).

3.2.2 Case 2: A single observer sends this message through a network of n independent channels in parallel

Two questions arise:

(i) Does the use of several channels enhance system performance in the sense of giving an improved probability of making the correct inference and/or a reduced probability of erasure?

(ii) Does it pay to use several observers, each with his own independent channel, rather than one?

Note that we do not discuss the loss of channels which might be used for other possibly more important messages.

If the single observer makes a correct interpretation of the message in the field the generating function corresponding to (143) is $(sg/x+\bar{s}+s\bar{g}x)^n$: otherwise it is $(s\bar{g}/x+\bar{s}+sgx)^n$. Thus, all information about the stochastic parallel information network is embodied in a function $h_n(x)$ given by

$$h_n(x) = c(sg/x+\bar{s}+s\bar{g}x)^n + \bar{c}(s\bar{g}/x+\bar{s}+sgx)^n. \tag{159}$$

In terms of the generating function $g_n(x)$ given by (143) we have

$$h_n(x) = c\, g_n(x|c=1) + \bar{c}\, g_n(x|c=0) \tag{160}$$

since $r = c\bar{g} + \bar{c}g$, $\bar{r} = cg + \bar{c}\bar{g}$ reduce to the required values when $c = 0$ and 1. The notation $g_n(x|c=c')$ means put $c = c'$ in $g_n(x)$.

We shall now give specific expressions for the basic quantities of interest in cases 1 and 2 for $n = 1,2,3,4$. Some numerical values will then be calculated and discussed briefly.

The quantities of interest are:

M_n Probability of erasure
N_n Probability of correct inference
PR   Conditional probability of correct inference, given no erasure
I_E The average gain in information.

We shall differentiate the single observer case by using a prime.

In the first place it is clear that M'_n is in form identical with M_n, r replacing g. It follows that M_n exceeds M'_n as long as

$$r\bar{r} > g\bar{g}.$$

Now

$$r\bar{r} - g\bar{g} = c\bar{c}(g-\bar{g})^2$$

and thus $M_n > M'_n$ for any c and g except $g = 0.5$, when they are equal. They are also equal for any g when $c = 0$ or 1. Thus, except in these rare circumstances, there is a higher probability of erasure of messages passed by several observers compared with that of a single message passed down the same number of channels.

It is more difficult to make such general statements about $N_n - N'_n$. The particular values are as follows:

$n = 1$

$M_1 = \bar{s}$, $M'_2 = \bar{s}$

$N_1 = sr$, $N'_1 = sr$

Generalized theory

$n = 2$

$$M_2 = \bar{s}^2 - 2s^2 r\bar{r}, \qquad M_2' = \bar{s}^2 + 2s^2 g\bar{g}$$
$$N_2 = 2s\bar{s}r + s^2 r^2, \qquad N_2' = 2s\bar{s}r + s^2(c\bar{g}^2 + \bar{c}g^2)$$

$n = 3$

$$M_3 = \bar{s}^3 + 6\bar{s}s^2 r\bar{r}, \qquad M_3' = \bar{s}^3 + 6\bar{s}s^2 g\bar{g}$$
$$N_3 = 3\bar{s}^2 sr + 3\bar{s}s^2 r^2 + s^3(r^3 + 3r^2\bar{r}),$$
$$N_3' = 3\bar{s}^2 sr + 3\bar{s}s^2(c\bar{g}^2 + \bar{c}g^2) + s^3(c\bar{g}^3 + \bar{c}g^3 + 3rg\bar{g})$$

$n = 4$

$$M_4 = \bar{s}^4 + 12\bar{s}^2 s^2 (r\bar{r})^2, \qquad M_4' = \bar{s}^4 + 12\bar{s}^2 s^2 g\bar{g} + 6s^4(g\bar{g})^2$$
$$N_4 = 4\bar{s}^3 sr + 6\bar{s}^2 s^2 r^2 + 4\bar{s}s^3(r^3 + 3r^2\bar{r}) + s^4(r^4 + 4r^3\bar{r})$$
$$N_4' = 4\bar{s}^3 sr + 6\bar{s}^2 s^2(c\bar{g}^2 + \bar{c}g^2) + 4\bar{s}s^3(c\bar{g}^3 + \bar{c}g^3 + 3g\bar{g}r) +$$
$$s^4(c\bar{g}^4 + \bar{c}g^4 + 4g\bar{g}(c\bar{g}^2 + \bar{c}g^2)).$$

After some algebra we find that

$$N_2 - N_2' = -c\bar{c}s^2 (\bar{g} - g)^2 \leq 0,$$

while

$$N_3 - N_3' = -c\bar{c}(\bar{g} - g)^2 s^2((1 - 2r)s + 3\bar{s}),$$

whose sign is not unequivocal.

Table 21 gives values of three quantities.

- PR_n: the probability that the fusion centre makes the correct inference, given that an inference is made.
- M_n: the probability of erasure, that is, insufficient information available for an inference to be made using the 'majority algorithm' described in Sections 3.1 and 3.2.
- I_E: the mean increase in information resulting from the messages.

The independent variables are, in order, p, c, s, q, n.

p: the probability that the message presented to the observer is the binary digit 1. The alternative is that it is 0, with probability $\bar{p} = 1 - p$. This is less restrictive than may seem. For example:

(a) the area searched contains less then two targets = 1; at least two = 0;

(b) the landing will take place between Calais and Boulogne = 1;

Table 21 — Comparison between one observer using n channels and n observers using one channel each

$p = 0.5$

		$c = 0.1$						$c = 0.9$					
		PR_n		M_n		I_E		PR_n		M_n		I_E	
	n	one	n	one	n	one	n	one	n	one	n	one	n
$s = 0.1$ $g = 0.1$	1	0.18	0.18	0.9	0.9	0.0222	0.0222	0.82	0.82	0.9	0.9	0.0222	0.0222
	2	0.1769	0.1750	0.8118	0.8130	0.0426	0.0429	0.8231	0.8251	0.8118	0.8130	0.0426	0.0429
	3	0.1739	0.1691	0.7339	0.7370	0.0615	0.0624	0.8261	0.8301	0.7339	0.7370	0.0615	0.0624
	4	0.1710	0.1650	0.6649	0.6705	0.0790	0.0808	0.8290	0.8350	0.6649	0.6705	0.0790	0.0808
$g = 0.5$	1	0.5	0.5	0.9	0.9	0	0	0.5	0.5	0.9	0.9	0	0
	2	0.5	0.5	0.815	0.815	0	0	0.5	0.5	0.815	0.815	0	0
	3	0.5	0.5	0.7425	0.7425	0	0	0.5	0.5	0.7425	0.7425	0	0
	4	0.5	0.5	0.6804	0.6804	0	0	0.5	0.5	0.6804	0.6804	0	0
$g = 0.9$	1	0.82	0.82	0.9	0.9	0.0222	0.0222	0.18	0.18	0.9	0.9	0.0222	0.0222
	2	0.8231	0.8250	0.8118	0.8130	0.0426	0.0429	0.1769	0.1750	0.8118	0.8130	0.0426	0.0429
	3	0.8267	0.8301	0.7339	0.7370	0.0615	0.0624	0.1739	0.1699	0.7339	0.7370	0.0615	0.0624
	4	0.8290	0.8350	0.6649	0.6705	0.0802	0.0808	0.1710	0.1650	0.6649	0.6705	0.0790	0.0808
$s = 0.5$ $g = 0.1$	1	0.18	0.18	0.5	0.5	0.1109	0.1109	0.82	0.82	0.5	0.5	0.1109	0.1109
	2	0.1596	0.1451	0.295	0.3238	0.1792	0.1887	0.8404	0.8548	0.295	0.3238	0.1792	0.1887
	3	0.1499	0.1182	0.1925	0.2357	0.2264	0.2521	0.8557	0.8818	0.1925	0.2357	0.2264	0.2521
	4	0.1332	0.0975	0.1330	0.1814	0.2607	0.3069	0.8668	0.9025	0.1330	0.1814	0.2607	0.9058
$g = 0.5$	1	0.5	0.5	0.5	0.5	0	0	0.5	0.5	0.5	0.5	0	0
	2	0.5	0.5	0.375	0.375	0	0	0.5	0.5	0.375	0.375	0	0
	3	0.5	0.5	0.3125	0.3125	0	0	0.5	0.5	0.3125	0.3125	0	0
	4	0.5	0.5	0.2734	0.2734	0	0	0.5	0.5	0.2734	0.2734	0	0
$g = 0.9$	1	0.82	0.82	0.5	0.5	0.1109	0.1109	0.18	0.18	0.5	0.5	0.1109	0.1109
	2	0.8404	0.8549	0.295	0.3238	0.1792	0.1887	0.1596	0.1451	0.295	0.3238	0.1792	0.1887
	3	0.8557	0.8818	0.1925	0.2957	0.2264	0.2521	0.1443	0.1182	0.1925	0.2357	0.2264	0.2521
	4	0.8668	0.8025	0.1330	0.1814	0.2607	0.3059	0.1332	0.0975	0.1330	0.1814	0.2607	0.3058
$s = 0.9$ $g = 0.1$	1	0.18	0.18	0.1	0.1	0.1996	0.1996	0.82	0.82	0.1	0.1	0.1996	0.1996
	2	0.1247	0.0781	0.1558	0.2491	0.2675	0.3147	0.8753	0.8219	0.1558	0.2491	0.2675	0.3147
	3	0.1214	0.0810	0.0447	0.0727	0.3090	0.3820	0.8786	0.9190	0.0447	0.0727	0.3090	0.3820
	4	0.1095	0.0449	0.0407	0.1002	0.3334	0.4588	0.8905	0.9557	0.0407	0.1002	0.3334	0.4588
$g = 0.5$	1	0.5	0.5	0.1	0.1	0	0	0.5	0.5	0.1	0.1	0	0
	2	0.5	0.5	0.415	0.415	0	0	0.5	0.5	0.415	0.415	0	0
	3	0.5	0.5	0.1225	0.1225	0	0	0.5	0.5	0.1245	0.1245	0	0
	4	0.5	0.5	0.2704	0.2704	0	0	0.5	0.5	0.2704	0.2704	0	0
$g = 0.9$	1	0.82	0.82	0.1	0.1	0.1996	0.1996	0.18	0.18	0.1	0.1	0.1996	0.1996
	2	0.8753	0.9219	0.1558	0.2491	0.2675	0.3147	0.1247	0.0781	0.1558	0.2491	0.2675	0.3147
	3	0.8786	0.9190	0.0447	0.0727	0.3090	0.3820	0.1214	0.0810	0.0447	0.0727	0.3090	0.3828
	4	0.8905	0.9537	0.0407	0.1002	0.3334	0.4588	0.1095	0.0449	0.0407	0.1002	0.3334	0.4588

Sec. 3.2] **Generalized theory** 81

elsewhere = 0.

c: The probability that the observer correctly identifies, or decodes, the message.
s: The probability that each channel has of being in a condition to transmit information.
g: The probability that each channel has of reversing the sense of the message sent.
n: The number of information channels available to the observers, $n = 1,2,3,4$.

In point of fact p is irrelevant to the calculation of PR and M_n, and enters only as a prior probability whose value is modified by the information (in the technical sense) received at the fusion centre.

For each value of n, two values are given. The first column refers to a *single* observer using n channels, the second to n independent observers each using an independent channel.

3.2.3 Comments on the tables
Comment (1): For fixed s and c and each n:

(a) $PR(g) = 1 - PR(\bar{g})$
(b) $M_n(g) = M_n(\bar{g})$ for both a single observer and many.
(c) $I_E(g) = I_E(\bar{g})$

Note that the arguments are introduced for emphasis. Of course, PR and the others depend on n, c, s as well as g.

(1b) is obvious theoretically since M_n in both cases is a function of $r\bar{r} = (c\bar{g} + \bar{c}g)(cg + \bar{c}\bar{g})$ which is symmetrical in g and $\bar{g}$. It is also symmetrical in c and $\bar{c}$, a reference to which will be made below.

(1a) We have $PR = N/(1 - M)$ for fixed n, which is omitted as a subscript. For (a) to be true

$$N(g)/(1 - M(g)) = 1 - N(\bar{g})/(1 - M(\bar{g})) = 1 - N(\bar{g})/(1 - M(g)) \text{ by (b)},$$

or

$$N(g) = 1 - M(g) - N(\bar{g}) = 1 - M(\bar{g}) - N(\bar{g}) = L(\bar{g}).$$

This is, by definition, true, and so (a) could have been predicted.

(1c) This symmetry is independent of the probabilities p of the message. By definition (see (147)) we have

$$I_E(g) = N(g)\log(N(g)) + L(g)\log(L(g)) - J(g)\log(J(g)) - K(g)\log(K(g))$$

where

$$J = \bar{p}N + pL, \qquad K = \bar{p}L + pN.$$

Since

$$N(\bar{g}) = L(g)$$

the first two terms interchange when g is replaced by $\bar{g}$. Also

$$J(\bar{g}) = \bar{p}N(\bar{g}) + pL(\bar{g}) = \bar{p}L(g) + pN(g) = K(g).$$

The result (c) follows.

Note that the symmetries (1a), (b), (c) hold also with respect to c for fixed s and g and each n. The proofs are similar. Similar proofs substantiate the results for a single observer.

Comment (2): There is another symmetry in c. We see that

(a) $PR(g,c) = PR(\bar{g},\bar{c})$

(b) $M_n(\bar{g},\bar{c}) = M_n(g,c)$

(c) $I_E(g,c) = I_E(\bar{g},\bar{c})$

These observations emphasize the fact that the model evaluates the performance of a bad observer with a noisy channel as highly as that of a good observer with a noise-free channel.

(2b) Replacing $\bar{c}$ by c as well as $\bar{g}$ by g leaves the value of $r\bar{r}$ unaltered. Hence 2(b) is to be expected.

2(a) If $PR(g,c) = PR(\bar{g},\bar{c})$,

$$\frac{N(g,c)}{L(g,c) + N(g,c)} = \frac{N(\overline{g,c})}{L(\overline{g,c}) + N(\overline{g,c})} = \frac{L(\overline{g,c})}{N(\overline{g,c}) + L(\overline{g,c})}.$$

But

$$r(c) = c\bar{g} + \bar{c}g, \quad \bar{r}(c) = cg + \bar{c}\bar{g}$$

so that

$$r(\bar{c}) = \bar{c}\bar{g} + cg = \bar{r}(c), \quad \bar{r}(\bar{c}) = \bar{c}g + c\bar{g} = r(c).$$

Thus, putting $\bar{c}$ for c in L exchanges $\bar{r}(c)$ for $r(\bar{c})$, and $r(c)$ for $\bar{r}(\bar{c})$. Thus

$L(g,\bar{c}) = N(g,c)$ and

$$PR(g,c) = \frac{L(g,\bar{c})}{L(g,\bar{c}) + N(g,c)} = \frac{N(g,\bar{c})}{L(g,c) + N(g,c)} = PR(\bar{g},\bar{c}),$$

quod erat demonstrandum.

(2c) Similar argument substantiates 2(c) theoretically.
In what follows the prime denotes the value corresponding to a single observer and n channels.

Comment (3): The tables are consistent with the proven fact that for fixed $n > 1$ and $g \neq \bar{g}$, $M_n > M'_n$ and that when $g = \bar{g} = 0.5$, $M_n = M'_n$.

Comment (4): The figures suggest that, uniformly with respect to g and for each $n > 1$,

(a) $PR' > PR$ for $g < \bar{g}$ and $c < \bar{c}$
$\qquad\qquad\qquad g > \bar{g}$ and $c > \bar{c}$
(b) $PR' = PR = 0.5$ for $g = \bar{g}$
$\quad PR' < PR$ for $g > \bar{g}, c > \bar{c}$
$\qquad\qquad\qquad g < \bar{g}, c < \bar{c}$.

This will be shown below to be true theoretically.

Comment (5): The figures suggest that

(a) As n increases, both I_E and I'_E increase. In other words, to add channels is to increase the mean gain in information.
(b) For fixed n, $I_E > I'_E$, that is, adding observers increases the mean gain in information. This accords with intuition.

Comment (6): As $s \to 1$ the phenomenon of oscillating M_n as well as M'_n is seen as n increases. This was explained above for M_n, and the same argument holds for M'_n. The tabular values of PR oscillate also in the same region as expected from earlier discussion; this is not seen to be the case for PR'. However, a finer grain tabulation does reveal the phenomenon, as might be expected.

Analytic examination of the sign of $PR' - PR$ is of interest. First, omitting subscripts, we have

$$PR' - PR = \frac{N'}{N' + L} - \frac{N}{N + L} = \frac{\Delta}{(N' + L')(N + L)}$$

where

$$\Delta \equiv \Delta(r,g) = L(r) N'(g) - L'(g) N(r) . \tag{161}$$

Thus, the sign of $PR' - PR$ is the sign of Δ.
To facilitate the analysis we recall that

$$N(x) = \binom{n}{1}\bar{s}^{n-1}sx + \binom{n}{2}\bar{s}^{n-2}s^2x^2 + \binom{n}{3}\bar{s}^{n-3}s^3(x^3 + \binom{3}{1}x^2\bar{x}) +$$

$$+ \binom{n}{4}\bar{s}^{n-4}s^4(x^4 + \binom{4}{1}x^3\bar{x}) + \ldots + \binom{n}{n}s^n(x^n + \binom{n}{1}x^{n-1}\bar{x} +$$

$$+ \ldots + \binom{n}{n_2}x^{n-n_2}\bar{x}^{n_2})$$

where $n_2 = n/2 - 1$ if n is even, and $(n-1)/2$ if n is odd. Also,

$$\begin{aligned}
L(r) &= N(\bar{r}) \\
N'(g) &= cN(\bar{g}) + \bar{c}N(g) \\
L'(g) &= cN(\bar{g}) + N(\bar{g})\bar{c} \\
r &= c\bar{g} + \bar{c}g \\
\bar{r} &= cg + \bar{c}\bar{g} = 1 - r .
\end{aligned}$$

Putting $g = \bar{g}$ we have $\bar{r} = g$, $r = g$, and

$$\begin{aligned}
\Delta(r,g) \to \Delta(g,g) &= L(g)N'(g) - L'(g)N(g) \\
&= N(\bar{g})(cN(g) + \bar{c}N(g)) - N(g)(cN(g) + \bar{c}N(\bar{g})) = 0 .
\end{aligned}$$

Thus $g - \bar{g}$ is a factor of $\Delta(r,g)$.
Similarly, putting $c = \bar{c}$ we have

$$r = c$$
$$\bar{r} = \bar{c}$$

and

$$\Delta(r,g) \to \Delta(c,g) = cN(c)(N(\bar{g}) + N(g)) - cN(c)(N(g) + N(\bar{g})) = 0$$

and $c - \bar{c}$ is also a factor of $\Delta(r,g)$.
Next put $c = 0$ and $\Delta(r,g) \to \Delta(g,g) = N(\bar{g})N(g) - N(g)N(\bar{g}) = 0$.
Similarly, if $c = 1$, $\Delta = 0$, and thus $c\bar{c}$ divides $\Delta(r,g)$. Hence we can write

Sec. 3.2] **Generalized theory** 85

$$\Delta(r,g) = c\bar{c}(c-\bar{c})(g-\bar{g}) E(r,g) . \qquad (162)$$

Finally, we need information about the sign of $E(r,g)$. This can be looked at for small c and g, and then also for c and g nearly equal to unity. Take the former case. First note that $N(0) = 0$ and $N(1) = 1 - \bar{s}^n$. Next

$$\frac{dN(x)}{dx} = \binom{n}{1}\bar{s}^{n-1}s + n(n-1)\bar{s}^{n-2}s^2 x + \binom{n}{3}\bar{s}^{n-3}s^3 (3x^2 + 2\binom{3}{1}x\bar{x} - \binom{3}{1}x^2) +$$

$$+ \binom{n}{4}\bar{s}^{n-4}s^4(4x^3 + \binom{4}{1}3x^2\bar{x} - \binom{4}{1}x^3) + \ldots + \binom{n}{n}s^n \left(nx^{n-1} + \right.$$

$$+ \binom{n}{1}(n-1)x^{n-2}\bar{x} - \binom{n}{1}x^{n-1} + \binom{n}{2}(n-2)x^{n-3}\bar{x}^2 - 2\binom{n}{2}x^{n-2}\bar{x} +$$

$$+ \ldots + \binom{n}{n_2}(n-n_2)x^{n-n_2-1}\bar{x}^{n_2} - n_2\binom{n}{n_2}x^{n-n_2}\bar{x}^{n_2-1}\right) .$$

Thus

$$(dN(x)/dx)_{x=0} = B_0 , \quad \text{where } B_0 = n\bar{s}^{n-1}s ,$$
$$(dN(x)/dx)_{x=1} = B_0 + B_1 , \text{where } B_1 = n(n-1)\bar{s}^{n-2}s^2 .$$

For small x we then have to the first order in x,

$$N(x) = B_0 x$$
$$N(\bar{x}) = N(1) - x(B_0 + B_1) .$$

Now let c and g be so small that c^2, g^2, cg and higher order terms can be neglected. To this order, $r = c + g$, $\bar{r} = 1 - c - g$,

$$L(r) = N(\bar{r}) = N(1) - r(B_0 + B_1)$$
$$N'(g) = cN(\bar{g}) + \bar{c}N(g) = cN(1) + B_0 g$$
$$L'(g) = cN(g) + \bar{c}N(\bar{g}) = N(1) - cN(1) - g(B_0 + B_1)$$
$$N(r) = B_0(c + g) .$$

To the same order

$$\Delta(r,g) = cN(1)(N(1) - B_0) = c(1 - \bar{s}^n)(1 - \bar{s}^n - n\bar{s}^{n-1}s) > 0 \quad \text{when } 0 < s < 1 .$$

From (162), to the same order,

$$\Delta(r,g) = cE(r,g) ,$$

and it can be concluded that $E(r,g) > 0$ over the range of practical interest and, moreover, that $E(r,g)$ contains $N(1) = 1 - \bar{s}^n$ as a factor. Note in particular that for $n = 2$

$$\Delta(r,g) = c\bar{c}(c-\bar{c})(g-\bar{g})^3 (1-\bar{s})^2 .$$

This is exact, and there may be a simple exact general formula for arbitrary n, only we have not yet found it if there is.

In conclusion we have shown that

$\Delta(r,g) > 0$ when $0 < c < \frac{1}{2}, 0 < g < \frac{1}{2}$, and $\frac{1}{2} < c < 1, \frac{1}{2} < g < 1$

$\Delta(r,g) = 0$ when $c = 0, \frac{1}{2}, 1$ and when $g = \frac{1}{2}$

$\Delta(r,g) < 0$ when $0 < c < \frac{1}{2}, \frac{1}{2} < g < 1$, and $\frac{1}{2} < c < 1, 0 < g < \frac{1}{2}$.

Thus, for fixed $0 < c < \frac{1}{2}$, $\Delta(r,g)$ increases from zero to a maximum as g increases from zero, returning to zero when $g = \frac{1}{2}$; it then decreases to a minimum as g increases from $\frac{1}{2}$ and returns to zero when $g = 1$. The behaviour for fixed g and variable c is similar.

The fact that when $g = \bar{g} = 0.5$, $PR = PR' = 0.5$ follows because in this case $r = 0.5$, $N(r) = L(r)$, and $PR = N(r)/(N(r) + L(r)) = 0.5$. Similarly for PR'.

4

Management of an information channel with a priority facility

4.1 INTRODUCTION TO 'HEAD-OF-THE-LINE' PRIORITY

The models and analysis presented in this chapter and the next are addressed to the problem of quantifying the degradation that may be imposed on a communication system by the abuse of priority. There is quite a long road to travel before the goal is reached, and the mathematical details will require patience on the part of those readers unacquainted with queuing theory, even though the methods are quite routine. Some technical notes are given in Appendix A. The practician concerned with the management of information flow may wish to jump straight to Sections 4.7.2, 4.8.3, 4.8.5 and 4.9, where the findings are presented and discussed.

The main principles and lessons can be extracted from a simple model. This consists of a single channel service (or queuing) system. Messages are supposed to arrive for onward transmission in a Poisson stream with constant rate λ requests per unit time. The lengths of messages are random variables with a common exponential distribution, parameter μ: μ is also a rate, and its reciprocal is the mean time required to transmit the message. It is also taken to be time-independent. Two further assumptions are made:

(i) the stream of requests and the service mechanism are statistically independent;
(ii) a state of statistical equilibrium reigns.

Roughly speaking, assumption (ii) means that the service is not swamped by the demand and that the system has been in operation long enough for the various probabilities of interest, for example the number of messages awaiting transmission, to be time-independent. The condition for this is that $\lambda < \mu$. In real life there will be periods when $\lambda > \mu$. In this case the probability that a finite number of messages awaits tends to zero. But there is little point in trying to understand chaos before coming to terms with order which itself, as we shall see, holds alarming prospects for non-priority messages.

To silence, at least temporarily, potential critics of the model's simplicity, we note that:

(i) Multiple transmission channels can be simulated roughly by appropriate multiplication of μ;
(ii) A 'loss' system can be analysed, albeit with some trouble. Such a system imposes a ceiling on the total number of messages allowed to wait. When the system is full, new messages are 'turned away'. This models management of the system by adjustment of demand. There are other methods.

Finally, let it be said that the model and assumptions are essentially those underlying the studies made by Erlang, in the early years of this century, of the Copenhagen telephone service. This work laid the foundations of modern queueing theory. A thorough discussion of the principles of priority queues is given in [14].

4.2 THE MECHANISM

The arrival stream will be assumed to comprise two parts. A fraction α of messages are assigned priority in the sense that, irrespective of other messages waiting, they are serviced as soon as the channel becomes free. We do not assume that they replace a non-priority message currently in service: that would be unrealistic in this context. This means that the mean arrival rate of priority messages is $\alpha\lambda$, and that of non-priority messages is $\bar{\alpha}\lambda$, where $\bar{\alpha} = 1 - \alpha$. In considering the waiting time processes of priority and non-priority messages, which is of central interest, it will be assumed that those messages which arrive when the channel is busy wait in a 'queue' and are subsequently serviced in order of arrival. It is convenient to think thus of two queues, one of priority messages which is serviced as long as it contains members, and a second of non-priority messages which can receive attention only when the other is empty.

4.3 STATISTICAL DESCRIPTION OF SYSTEM STATE

The first task is to describe the system state. This is a vector random variable with three components (X,Y,Z). X and Y are, respectively, the numbers of messages in the priority and non-priority queues, including the message in service. $Z = 1$ when a priority message is being transmitted, and $Z = 0$ when a non-priority message is being transmitted. The value of Z when $X = Y = 0$ is immaterial. Let

$$a_{m,n} = P(X = m, Y = n, Z = 0) \quad (m \geq 1, n \geq 0)$$
$$b_{m,n} = P(X = m, Y = n, Z = 0) \quad (n \geq 1, m \geq 0)$$
(163)

Note: When there is no risk of confusion the subscripts will not be separated by a comma.

Statistical description of system state

To an onlooker concerned only with the total number of messages this is a stationary $M/M/1$ (see Appendix A) system, and it is known that

$$p_n = P(X+Y = n) = r^n(1-r), \qquad (n \geq 0) \tag{164}$$

where

$$r = \lambda/\mu .$$

Thus,

$$\sum_{m=1}^{n} a_{m,n-m} + \sum_{m=0}^{n-1} b_{m,n-m} = p_n . \tag{165}$$

The difference equations satisfied by the probabilities can be obtained by a technique called the 'method of balance', or otherwise, according to taste. The method of balance is the expression of a conservation principle which must hold for a system of this kind in statistical equilibrium; the details can be found in reasonable elementary texts on queuing theory. We have:

$$a_{1,0} + b_{0,1} = p_1 = rp_0 \tag{166}$$
$$(1+r)\, a_{1,0} = a_{2,0} + b_{1,1} + \alpha\, rp_0 \tag{167}$$
$$(1+r)\, b_{0,1} = a_{1,1} + b_{0,2} + \bar{\alpha}\, rp_0 . \tag{168}$$

More generally, for a total of $n>1$ messages waiting, including that being transmitted.

$$(1+r)\, a_{n0} = a_{n+1\,0} + b_{n\,1} + \alpha r\, a_{n-1\,0} \qquad (n \geq 2)$$
$$(1+r)\, a_{n-1\,1} = a_{n\,1} + b_{n-1\,2} + \alpha r\, a_{n-2\,1} + \bar{\alpha} r\, a_{n-1\,0}$$
$$(1+r)\, a_{n-2\,2} = a_{n-1\,2} + b_{n-2\,3} + \alpha r\, a_{n-3\,2} + \bar{\alpha} r\, a_{n-2\,1} \qquad (n \geq 3)$$
$$\cdots\cdots\cdots\cdots\cdots\cdots\cdots\cdots\cdots\cdots\cdots\cdots$$
$$(1+r)\, a_{2\,n-2} = a_{3\,n-2} + b_{2\,n-1} + \alpha r\, a_{1\,n-2} + \bar{\alpha} r\, a_{2\,n-3} \tag{169}$$
$$(1+r)\, a_{1\,n-1} = a_{2\,n-1} + b_{1n} + \bar{\alpha} r\, a_{1\,n-2}$$
$$(1+r)\, b_{0\,n} = a_{1\,n} + b_{0\,n+1} + \bar{\alpha} r\, b_{0\,n-1} \qquad (n \geq 2) \tag{170}$$
$$(1+r)\, b_{1\,n-1} = \alpha r\, b_{0\,n-1} + \bar{\alpha} r\, b_{1\,n-2}$$
$$(1+r)\, b_{2\,n-2} = \alpha r\, b_{1\,n-2} + \bar{\alpha} r\, b_{2\,n-3} \qquad (n \geq 3)$$
$$\cdots\cdots\cdots\cdots\cdots\cdots\cdots\cdots\cdots\cdots\cdots\cdots$$
$$(1+r)\, b_{n-2\,2} = \alpha r\, b_{n-3\,2} + \bar{\alpha} r\, b_{n-2\,1}$$
$$(1+r)\, b_{n-1\,1} = \alpha r\, b_{n-2\,1} . \qquad (n \geq 2) \tag{171}$$

Summation of these equations and use of (165) yields

$$(1+r)\, p_n = p_{n+1} + r\, p_{n-1} , \qquad (n \geq 1) \tag{172}$$

the familiar difference equation satisfied by the $M/M/1$ equilibrium state probabilities with solution given by (164)

4.4 NON-PRIORITY SYSTEM TIME: GENERAL DISCUSSION

Before discussing the solution of the set (166–171), it is worthwhile examining briefly the plight of a non-priority message. For this purpose let us consider a random variable called here 'system time', the sum of queueing time and transmission time, often called 'waiting time including service' in queueing texts. The system time of a priority message is reasonably uncomplicated and is, in general, the sum of the residual transmission times of the priority messages ahead and own transmission time. Thus we are dealing with weighted sums of random variables whose densities are autoconvolutions of the basic transmission time density, $b(t) = \mu \exp(-\mu t)$, with weights derived from appropriate combinations of the solutions of (166–171).

The case of a non-priority message is different, and can be illuminated by designating special epochs in the evolution of the process as *non-priority service opportunities*. Such an epoch occurs when a message has just been transmitted and the priority queue is empty. This means that the next message in the non-priority queue can be processed.

Suppose that such a service opportunity has just occurred and that a non-priority message has begun transmission. Let X be the time to the next non-priority service opportunity. By definition the priority queue was empty at the start. If no priority message arrives during the transmission time S, $X = S$: if one priority message arrives X is equal to S plus the transmission time B_1 of the priority arrival and of all the priority arrivals which take place until the priority queue is next empty and the message in service has completed transmission. B_1 is called a 'busy period' in the queueing literature. If n priority arrivals occur during S, $X = S + B_n$, where B_n is the sum of the busy periods initiated by each of the n priority arrivals.

Let B_1 have probability density function $k(t)$ and denote by $k_n(t)$ the n-fold autoconvolution of $k(t)$. Then, conditional on there being n priority arrivals during S, the probability density function of X is $b(t)*k_n(t)$, where * denotes convolution of the functions it separates. Given that the Poisson arrival process of priority customers has mean rate $\alpha\lambda$ we then find for the probability density function $f(t)$ of X the expression

$$f(t) = \mu e^{-\alpha\lambda t - \mu t} + e^{-\alpha\lambda t} \sum_{n \geq 1} \frac{(\alpha\lambda t)^n}{n!} \mu e^{-\mu t} * k_n(t) \ .$$

Let

$$\phi(z) = \int_0^\infty e^{-zt} f(t) \, dt$$

be the Laplace transform of $f(t)$ and $\kappa(z)$ that of $k_1(t)$. Finally let

Sec. 4.4] Non-priority system time: general discussion

$$Z = z + \mu + \alpha\lambda .$$

Then

$$\phi(z) = \frac{\mu}{Z} \sum_{n \geq 0} \left(\frac{\alpha\lambda}{Z}\right)^n \frac{\Gamma(n+1)}{n!}$$

$$= \frac{\mu}{Z - \alpha\lambda\, \kappa(z)} . \qquad (173)$$

It is known that, provided that $\alpha\lambda < \mu$,

$$\kappa(z) = (Z - R)/(2\alpha\lambda) \qquad (174)$$

where

$$R^2 = Z^2 - 4\alpha\lambda\mu . \qquad (175)$$

Thus $\kappa(z)$ is the smaller zero of the quadratic $\alpha\lambda x^2 - Zx + \mu$ and, by substitution in (173) it follows in this case that $\phi(z) \equiv \kappa(z)$. This means that X and B_1 are statistically identical, but we shall for the time being preserve the distinction to clarify explanation.

Note that when $\alpha\lambda < \mu$, B_1 is a proper random variable (finite with probability one). This means that the system as a whole could be operated with $\lambda > \mu$, which would mean that, overall, it would be in disequilibrium and that the number of messages waiting would grow indefinitely. But if, at the same time, $\alpha\lambda < \mu$, the priority queue would be in equilibrium and priority messages would have finite waiting times, while the queue of the non-priority messages would grow indefinitely, giving rise to the increasing likelihood of their never being transmitted at all. This is a possible consequence of the abuse of priority to which allusion was made at the beginning.

Although X is only a component of the system time of non-priority message it is of interest to calculate $E[X]$ and $\text{Var}[X]$. These can be deduced directly from $\phi(z)$. We have

$$\ln \phi(z) = \ln \mu - \ln(Z - \alpha\lambda\kappa(z)) .$$

Thus

$$\frac{\phi'(z)}{\phi(z)} = -\frac{(1 - \alpha\lambda\kappa'(z))}{Z - \alpha\lambda\kappa(z)}$$

and

$$\frac{\phi''(z)}{\phi(z)} - \left[\frac{\phi'(z)}{\phi(z)}\right]^2 = \frac{\alpha\lambda\kappa''(z)}{Z - \alpha\lambda\kappa(z)} + \left[\frac{1 - \alpha\lambda\kappa'(z)}{Z - \alpha\lambda\kappa(z)}\right]^2.$$

Since when $\alpha\lambda < \mu$, $\phi(0) = 1$,

$$\phi'(0) = -E[X] , \quad \frac{\phi''(0)}{\phi(0)} - \left[\frac{\phi'(0)}{\phi(0)}\right]^2 = \text{Var}[X] ,$$

we have

$$E[X] = \frac{1}{\mu(1-\alpha r)} , \quad \text{Var}[X] = \frac{1+\alpha r}{\mu^2(1-\alpha r)^3} , \qquad (176)$$

where $r = \lambda/\mu$. Notice that $\text{Var}[X]$ is inversely proportional to $(1-\alpha r)^3$ which is worse than exponential and disastrous as $\alpha r \to 1$.

The corresponding statistics of X for priority messages are $1/\mu$ and $1/\mu^2$, since in this case X is statistically identical with the service time, and so, although X is but a component of system time, it is clear that the priority system can disadvantage non-priority messages to such an extent that they never get transmitted at all.

4.5 GENERAL FORM OF PROBABILITY DENSITIES OF SYSTEM TIME

A full treatment of the system time problem calls for solution of (166–171) which will be treated later.

Let $g_p(t)$, $g_{np}(t)$ denote respectively the system time probability density functions of priority and non-priority messages. Consider first priority messages and write

$$A_m = \sum_{n \geq 0} a_{mn} \quad (m \geq 1) , \qquad (177)$$

$$B_m = \sum_{n \geq 1} b_{mn} \quad (m \geq 0)$$

and

$$f_n(t) = f_{n-1}(t) * f(t) ,$$

so that f_n is the n-fold autoconvolution of f.

Then

$$g_p(t) = \mu e^{-\mu t} \left[p_0 + \sum_{m \geq 1} (A_m + B_{m-1})(\mu t)^m/m! \right]. \quad (178)$$

The explanation is as follows.

For a priority message to arrive and find $m-1$ priority messages ahead in the queue, and one more in transmission, the probability is

$$\sum_{n \geq 0} a_{mn} = A_m.$$

The time required to transmit the residue of the message in transmission, the next $m-1$ messages, and then the message which has just arrived, has a probability density function which is the $(m+1)$-fold autoconvolution of $\mu \exp(-\mu t)$, namely

$$\mu e^{-\mu t}(\mu t)^m/m!.$$

Similarly for a priority message to arrive and find $m (\geq 0)$ priority messages waiting and a non-priority message being transmitted has probability

$$\sum_{n \geq 1} b_{mn} = B_m.$$

The wait is the residual transmission time of the current message in transmission plus the sum of the transmission times of the m priority messages already present plus own transmission time; again the $(m+1)$-fold autoconvolution of $\mu \exp(-\mu t)$ is needed. This explains the terms under the summation sign in (178). The remaining term caters for the case where the new message finds an empty system.

For a non-priority message the situation is a little more complicated. The result is

$$g_{np}(t) = \mu p_0 e^{-\mu t} + \sum_{m \geq 1} a_{m0} k_m(t) * \mu e^{-\mu t} + \sum_{n \geq 1} b_{0n} f_n(t) * \mu e^{-\mu t}$$

$$+ \sum_{m \geq 1} \sum_{n \geq 1} a_{mn} k_m(t) * f_n(t) * \mu e^{-\mu t}$$

$$+ \sum_{m \geq 1} \sum_{n \geq 1} b_{mn} k_{m+1} * f_{n-1}(t) * \mu e^{-\mu t}. \quad (179)$$

The case of an empty system is dealt with as above. If the arriving message finds no message ahead in the non-priority queue and $m(\geqslant 1)$ messages in the priority part of the system (probability a_{m0}) it must await the clearance of the m and any other priority messages that arrive meanwhile. The associated probability density function is $k_m(t)$ and this must be convolved with $\mu \exp(-\mu t)$ to give system time. This accounts for the first sum in (179). If the arriving message finds $n(\geqslant 0)$ ahead in the non-priority *queue*, one in transmission, and no priority message present, it must wait effectively through n intervals X between service opportunities before being transmitted. Note that lack of memory of the service time distribution of the message being transmitted permits the n-fold convolution entailed to be written $f_n(t)$. This accounts for the second term in (179). The last two terms correspond to the case where the newly arrived non-priority message finds at least one message of either kind present. Suppose that there are m priority, and n priority, messages. If a priority message is being transmitted, the time to the next service opportunity is a busy period beginning with m and has probability density function $k_m(t)$. The further wait is the time required to service the n non-priority messages ahead, and to this must be added own transmission time, giving probability density function

$$k_m(t) * f_n(t) * \mu \exp(-\mu t) \ .$$

If a non-priority message is being transmitted on arrival, the time until the service next becomes available for a non-priority messaage is a busy period beginning with $m+1$ (that is the one in transmission and the m priority messages waiting), while the remaining wait is the sum of the $n+1$ interservice-opportunity intervals X and, finally, own transmission time. This completes the explanation of (179). We shall return to this in Section 4.8.4.

4.6 PROCESSING OF EQUATIONS (166–171) AND THE EXPLICIT FORM OF $g_p(t)$

Next we turn to the problem of processing the set of difference equations (166–171) for the state probabilities $a_{m,n-m}$ and $b_{m,n-m}$, with a view to finding $g_p(t)$.

To find $g_p(t)$ we must calculate A_m and B_{n-1}. It is in fact straightforward to construct the following difference equations for these quantities:

$$(1 + \alpha r) A_1 = A_2 + B_1 + \alpha r p_0 \tag{180}$$

$$(1 + \alpha r) A_2 = A_3 + B_2 + \alpha r A_1 \tag{181}$$

$$\dots\dots\dots\dots$$

$$(1 + \alpha r) A_m = A_{m+1} + B_m + \alpha r A_{m-1}, \ (m \geqslant 2) \ . \tag{182}$$

Since the form (182) contains A_1 when $m=2$, it follows that the solution found for A_m ($m \geqslant 2$) holds also for $m = 1$, and substitution into (180) will determine a constant.

In addition

$$(1+r) B_0 = A_1 - a_{10} + B_0 - b_{01} + \bar{\alpha} r p_0 + \bar{\alpha} r B_0$$

from which, using (166),

$$\alpha r B_0 = A_1 - \alpha p_1 . \tag{183}$$

For $n \geq 1$ we have also

$$(1 + \alpha r) B_n = \alpha r B_{n-1} \tag{184}$$

so that

$$B_n = \sigma^n B_0 \tag{185}$$

where

$$\sigma = \alpha r/(1 + \alpha r) . \tag{186}$$

The solution of (182) is of the form

$$A_m = K_1 \xi_1^m + K_2 \xi_2^m$$

plus any particular solution, u_m, which satisfies (182), given B_m. ξ_1 and ξ_2 are the roots of $x^2 - (1 + \alpha r)x + \alpha r = 0$, that is

$$\xi_1, \xi_2 = \frac{(1 + \alpha r) \pm (1 - \alpha r)}{2} .$$

The larger root is $\xi_1 = 1$, and the smaller, $\xi_2 = \alpha r (<1)$. Since ΣA_m must be finite (it is a probability) the ξ_1 term has to be excluded from the solution. Thus

$$A_m = K(\alpha r)^m + u_m$$

where

$$f(E) u_m = B_m = B_0 \sigma^m$$

and

$$f(E) = 1 + \alpha r - E - \alpha r/E .$$

E is the usual shift operator which has the effect $E^n u_m = u_{m+n}$ for positive and negative n.

To find u_m, consider

$$f(E) \sigma^m = \sigma^{m-1}(\alpha r - \sigma^2 - \alpha r) = -\sigma^{m+1}.$$

Thus

$$-\sigma^{m-1} = \frac{1}{f(E)} \sigma^m,$$

and the solution sought is

$$A_m = K(\alpha r)^m - B_0 \sigma^{m-1},$$

wherein K and B_0 have still to be found.

Since the above solution has to be valid for $m = 1$ we obtain by substitution into (180):

$$K = \frac{B_0}{\sigma} + p_0.$$

Thus

$$A_m = B_0[(1+\alpha r)(\alpha r)^{m-1} - \sigma^{m-1}] + p_0(\alpha r)^m. \tag{187}$$

It remains to find B_0, and this can be done by normalization or by using (183); pursuing the path of normalization we find that

$$1 = p_0 + \sum_{m=1}^{\infty}\sum_{n=0}^{\infty} a_{mn} + \sum_{m=0}^{\infty}\sum_{n=1}^{\infty} b_{mn} = p_0 + \sum_{m \geq 0} B_m + \sum_{m \geq 1} A_m.$$

After some calculation we get

$$B_0 = \bar{\alpha} r/(1+\alpha r) = (\bar{\alpha}/\alpha)\sigma, \tag{188}$$

and accordingly

$$A_m = \left(\frac{\bar{\alpha}}{\alpha}\right)[(\alpha r)^m - \sigma^m] + p_0(\alpha r)^m \qquad (m \geq 1) \tag{189}$$

$$B_n = \left(\frac{\bar{\alpha}}{\alpha}\right) \sigma^{n+1}, \qquad (n \geq 0)$$

from which we see that

$$A_m + B_{m-1} = \left(\frac{\bar{\alpha}}{\alpha} + p_o\right)(\alpha r)^m . \tag{190}$$

Substituting into (178) and summing gives finally

$$g_p(t) = \mu\left[\left(\frac{1-\alpha r}{\alpha}\right) e^{-\mu(1-\alpha r)t} - \left(\frac{\bar{\alpha}}{\alpha}\right) e^{-\mu t}\right] . \tag{191}$$

A check on this is provided by putting $\alpha = 1$, so that all traffic is treated in the same way. In this case the system reduces to the standard $M/M/1$ system for which system time is known to have density

$$\mu(1-r) e^{-\mu(1-r)t} . \tag{192}$$

It is readily seen that (191) does indeed reduce to this form.

4.7 SOME STATISTICS OF SYSTEM TIME AND SYSTEM STATE

4.7.1 Formulae and the use of Little's Result

We can now establish the effect of a priority system for priority messages in terms of system time, and we can do the same for the less fortunate non-priority messages. Writing W_p for priority system time, we get

$$E[W_p] = \frac{(1+\bar{\alpha}r)}{\mu(1-\alpha r)},$$

$$\text{Var}[W_p] = \frac{1+2\bar{\alpha}r - r^2(1-\alpha^2)}{\mu^2(1-\alpha r)^2} . \tag{193}$$

Little's formula (see Appendix A) applied to the priority system gives for the mean of the number N_p of priority messages waiting,

$$E[N_p] = \alpha\lambda\, E[W_p] = \frac{\alpha r(1 + \bar{\alpha}r)}{(1 - \alpha r)}. \tag{194}$$

Since the mean number present in the total system is $r/(1-r)$, we find for the mean number $E[N_{np}]$ of messages in the non-priority system

$$E[N_{np}] = \frac{r}{1-r} - E[N_p] = \frac{\bar{\alpha}r(1 - \alpha r + \alpha r^2)}{(1-r)(1-\alpha r)}. \tag{195}$$

Using Little's formula again, we get for the mean system time $E[W_{np}]$ of non-priority messages,

$$E[W_{np}] = E[N_{np}]/\bar{\alpha}\lambda. \tag{196}$$

Some numerical values are given in Table 22.†.

† [Note: the validity of applying Little's formula to the priority subsystem alone can be verified by direct calculation. We observe that

$$P[N_p = 0] = p_0 + B_0, \quad P[N_p = n] = A_n + B_n \ (n \geq 1).$$

It is easily verified that

$$p_0 + B_0 + \sum_{n \geq 1}(A_n + B_n) = 1$$

(this is indeed how the constants were determined) and that

$$E[N_p] = \sum_{n \geq 1} n(A_n + B_n)$$

does have the value given by (194). Moreover,

$$E[N_p^2] = \sum_{n \geq 1} n^2(A_n + B_n) = \frac{r(1+\alpha r)}{(1-\alpha r)^2} - \bar{\alpha}r(1 + 2\alpha r),$$

and consequently

$$\text{Var}[N_p] = \frac{r}{(1-\alpha r)^2} - \bar{\alpha}r\left(1 + \alpha r + \frac{\alpha^2 r^3}{(1-\alpha r)^2}\right). \tag{197}$$

This confirms the validity of the simple approach to finding mean waiting time for the non-priority stream by means of Little's formula. No such simple approach is available for the variances.]

Table 22 — Mean equilibrium system times of priority and non-priority streams as a function of traffic intensity r and priority ratio α

α	Priority $r = 0.1$	Non-priority	α	Priority $r = 0.5$	Non-priority
0.1	1.1010	1.1122	0.1	1.5263	2.0526
0.2	1.1020	1.1133	0.2	1.5555	2.1111
0.3	1.1030	1.1145	0.3	1.5882	2.1765
0.4	1.1042	1.1157	0.4	1.6250	2.25
0.5	1.1033	1.1169	0.5	1.6666	2.3333
0.6	1.1064	1.1182	0.6	1.7143	2.4286
0.7	1.1075	1.1195	0.7	1.7692	2.5385
0.8	1.1087	1.1208	0.8	1.8333	2.6667
0.9	1.1098	1.1221	0.9	1.9091	2.8182

α	Priority $r = 0.7$	Non-priority	α	Priority $r = 0.9$	Non-priority
0.1	1.7527	3.5089	0.1	1.9890	10.8901
0.2	1.8139	3.7132	0.2	2.0976	11.9756
0.3	1.8361	3.9536	0.3	2.2329	13.3288
0.4	1.9722	4.2407	0.4	2.4062	15.0625
0.5	2.0769	4.5897	0.5	2.6364	17.3636
0.6	2.2069	5.0229	0.6	2.9565	20.5652
0.7	2.3725	5.5752	0.7	3.4324	25.3243
0.8	2.5909	6.3030	0.8	4.2143	33.1428
0.9	2.8919	7.3063	0.9	5.7368	48.3684

α	Priority $r = 0.95$	Non-priority	α	Priority $r = 0.99$	Non-priority
0.1	2.0497	21.9945	0.1	2.0988	110.8779
0.2	2.1728	24.4567	0.2	2.2344	124.4414
0.3	2.3287	27.5734	0.3	2.4082	141.8250
0.4	2.5323	31.6452	0.4	2.6391	164.9073
0.5	2.8095	37.1905	0.5	2.9604	197.0396
0.6	3.2093	45.1860	0.6	3.4384	244.0424
0.7	3.8358	57.7164	0.7	4.2248	323.4756
0.8	4.9583	80.1667	0.8	5.7596	476.9615
0.9	7.5517	132.0345	0.9	10.0826	909.2569

Note: the time unit is $1/\mu$, the mean time required to process a message.

4.7.2 Numerical values and comments

Traffic intensity is measured by r, the ratio of mean arrival rate to mean service rate. r also has an interpretation as the probability that the service is busy or, in this context, that messages are waiting and being transmitted. It is a convention to describe the region $0 < r \leq 0.5$ as 'light traffic', $0.5 < r \leq 0.8$ as 'medium traffic', and $0.8 < r < 1$ as 'heavy traffic'. Table 22 shows what is intuitively obvious, that under light traffic conditions the assignment of any priority ratio α makes little difference, but as traffic becomes heavier the difference between mean system times becomes increasingly significant. At a level of exceptionally heavy traffic ($r = 0.99$) the assignment of priority to 10% of the traffic leads to a 55-fold increase, while if priority is given to 90% of the traffic the difference is 90-fold. The mean system time of non-priority traffic seems quite intolerable and indicates that measures are needed to control traffic intensity, either by reduction of demand, for example application of a severe criterion for the acceptance of messages, or by increase of mean processing rate, for example by introducing one or more additional communication channels.

Four more points are worth making:

(i) To compare the effect of the priority system with one in which all messages are treated alike we note that in such a case the mean system time is $1/\mu(1-r)$. The general observation is that non-priority messages always have to wait longer than this, while the priority class always waits less, on average.
(ii) Increasing the proportion of priority messages always increases their system time, as well as that of the non-priority class, though less severely.
(iii) Increases in traffic intensity always increase mean system time for a given value of α.
(iv) As we shall see, the system time variance of non-priority messages shows up even worse by comparison with that of priority messages.

4.8 NON-PRIORITY SYSTEM STATE AND THE FORM OF $g_{np}(t)$

We shall now address the task of obtaining more information about non-priority system time. We shall obtain the Laplace transform $\gamma_{np}(z)$ of the probability density function $g_{np}(t)$ which means that all the moments of W_{np} can be calculated in principle. In particular we shall obtain an expression for $\text{Var}[W_{np}]$ to be compared numerically with $\text{Var}[W_p]$ in Table 24. The analysis also permits a direct evaluation of $E[N_{np}]$, which can be compared with (195) for checking, and $\text{Var}[N_{np}]$.

4.8.1 Generating function of state probabilities

The analysis is based on the introduction of the generating functions

$$A_m(x) = \sum_{n \geq 0} a_{mn} x^n \quad (m \geq 1)$$

$$B_m(x) = \sum_{n \geq 1} b_{mn} x^n \quad (m \geq 0) .$$

(198)

Thus A_m and B_n calculated in Section 4.6 are respectively $A_m(1)$ and $B_n(1)$. To get $B_0(x)$ we start in the usual way with the set:

$$\begin{aligned} x \quad & (1+r) b_{01} = a_{11} + b_{02} + \bar{\alpha} r p_0 \\ x^2 \quad & (1+r) b_{02} = a_{12} + b_{03} + \bar{\alpha} r b_{01} \\ x^3 \quad & (1+r) b_{03} = a_{13} + b_{04} + \bar{\alpha} r b_{02} \end{aligned}$$
. . . .

Multiplying by the power of x indicated in the margin and adding gives

$$(1+r) B_0(x) = A_1(x) - a_{10} + \frac{1}{x}(B_0(x) - b_{01}x) - \bar{\alpha} r p_0 x + \bar{\alpha} r x B_0(x)$$

which can be arranged into

$$\left(1 + r - \frac{1}{x} - \bar{\alpha}rx\right) B_0(x) = A_1(x) - p_1 + \bar{\alpha}xp_1 \tag{199}$$

after using (166).
Similarly, from the set

$$(1+r)\, b_{m1} = \alpha r\, b_{m-1\,1}$$
$$(1+r)\, b_{m2} = \alpha r\, b_{m-1\,2} + \bar{\alpha} r\, b_{m1}$$
$$(1+r)\, b_{m3} = \alpha r\, b_{m-1\,3} + \bar{\alpha} r\, b_{m2} \qquad (m \geqslant 1)$$

and so on, can be derived

$$B_m(x) = \sigma(x)\, B_{m-1}(x) \tag{200}$$

where

$$\sigma(x) = \alpha r/(1 + r - \bar{\alpha}rx) \tag{201}$$

leading to

$$B_m(x) = B_0(x)\,[\sigma(x)]^m \qquad (m \geqslant 0). \tag{202}$$

Similar treatment yields expressions for $A_m(x)$.
Thus

$$(1 + r - \bar{\alpha}rx)\, A_1(x) = A_2(x) + \frac{1}{x} B_1(x) + \alpha r p_0 \tag{203}$$

and

$$(1 + r - \bar{\alpha}rx)\, A_m(x) = A_{m+1}(x) + \alpha r\, A_{m-1}(x) + \frac{1}{x} B_m(x) \qquad (m \geqslant 2). \tag{204}$$

Since $A_1(x)$ enters (204) when $m = 2$, the general form of solution found holds also for $m = 1$, and then (203) provides a constraint from which a constant can be determined. As was the case for A_m, the solution of (204) is of the form

$$A_m(x) = K_1\, \eta_1^m + K_2\, \eta_2^m + \text{a particular solution},$$

where η_1 and η_2 are the roots of

$$y^2 - (1+r-\bar{\alpha}rx) y + \alpha r = 0 , \qquad (205)$$

namely

$$\eta_1, \eta_2 = \tfrac{1}{2}(X \pm R) \qquad (206)$$

where

$$\begin{aligned} X &= 1+r-\bar{\alpha}rx , \\ R^2 &= X^2 - 4\alpha r . \end{aligned} \qquad (207)$$

Since

$$(\eta_1 - 1)(1 - \eta_2) = \bar{\alpha}r(1-x)$$

is positive for $0 < x < 1$ (required to ensure convergence) it follows that the larger root is ineligible to form part of the solution. Thus we write

$$A_m(x) = K \eta^m(x) + \frac{B_0(x)}{x} \frac{1}{f(E)} \sigma^m(x)$$

where

$$f(E) = 1 + r - \bar{\alpha}rx - E - \frac{\alpha r}{E}$$

and $\eta(x)$ is the smaller root. K is independent of m. Since

$$f(E) \sigma^m(x) = \sigma^{m-1}(x) (\alpha r - \sigma^2(x) - \alpha r) = -\sigma^{m+1}(x)$$

we have

$$A_m(x) = K \eta^m(x) - \frac{B_0(x)}{x} \sigma^{m-1}(x) . \qquad (208)$$

If this is to satisfy (203) and (202) we get

$$K = p_0 + \frac{B_0(x)}{x\sigma(x)} \qquad (209)$$

Sec. 4.8] **Non-priority system state and the form of $g_{np}(t)$**

so that

$$A_m(x) = p_0 \, \eta^m(x) + \frac{B_0(x)}{x\sigma(x)} (\eta^m(x) - \sigma^m(x)) , \quad (m \geq 1) . \tag{210}$$

To complete the solution $B_0(x)$ is needed. This can be obtained from (199). We get

$$B_0(x) = \frac{x\sigma(x) \, p_0(\eta(x) - r(1 - \bar{\alpha}x))}{\alpha r x - \eta(x)} . \tag{211}$$

As a check this can be shown to reduce when $x = 1$ to (188). The operation is a little delicate since numerator and denominator contain $1 - x$ as a factor.

4.8.2 Non-priority system state

Before dealing with W_{np} let us consider N_{np}, in particular with a view to calculating $\text{Var}[N_{np}]$ and checking $E[N_{np}]$ as given in (194). The generating function

$$C(x) = E[x^{N_{np}}] \tag{212}$$

is easily seen to be given by

$$C(x) = p_0 + \sum_{m \geq 1} [A_m(x) + B_{m-1}(x)]$$

$$= p_0 + \frac{B_0(x)}{x} \left[\frac{x-1}{1-\sigma(x)} + \frac{\eta(x)}{\sigma(x)(1-\eta(x))} \right] . \tag{213}$$

As usual we can calculate the moments from

$$E[N_{np}] = \left[\frac{dC(x)}{dx} \right]_{x=1} ,$$

$$E[N_{np}(N_{np}-1)] = \left[\frac{d^2 C(x)}{dx^2} \right]_{x=1} . \tag{214}$$

The direct approach is very laborious and error-prone. A slight gain in manageability can be obtained by putting

$$X = 2(\alpha r)^{\frac{1}{2}} \cosh u = 1 + r - \bar{\alpha}rx , \tag{215}$$

in which case

$$R(x) = 2(\alpha r)^{\frac{1}{2}} \sinh u ,$$
$$\eta(x) = (\alpha r)^{\frac{1}{2}} e^{-u} ,$$
$$\sigma(x) = (\alpha r)^{\frac{1}{2}}/(2 \cosh u) ,$$

$$B_0(x) = \frac{p_0}{2(\alpha r)^{\frac{1}{2}}} \left[\frac{1 + r - 2(\alpha r)^{\frac{1}{2}} \cosh u}{\cosh u (1 - (r/\alpha)^{\frac{1}{2}} e^{-u})} \right] . \tag{216}$$

The gain is that the common factor $1 - x$ in the numerator and denominator of (211) can be cancelled explicitly when $B_0(x)$ is thus expressed in exponential form.

After further manipulations we arrive at the formula

$$C(x) = p_0(C_1(x) + C_2(x))$$

where

$$C_1(x) = \frac{1}{1 - (r/\alpha)^{\frac{1}{2}} e^{-u}} ,$$

$$C_2(x) = \frac{(e^u - (\alpha r)^{\frac{1}{2}})(e^{-u} - (\alpha r)^{\frac{1}{2}})}{(\alpha r)^{\frac{1}{2}}(2 \cosh u - (\alpha r)^{\frac{1}{2}})(1 - (r/\alpha)^{\frac{1}{2}} e^{-u})} . \tag{217}$$

The value u_1 of u corresponding to $x = 1$ is given by $e^{-u_1} = (\alpha r)^{\frac{1}{2}}$. This shows readily that $C_2(1) = 0$, $C_1(1) = (1 - r)^{-1}$ giving $C(1) = 1$, as should be the case.

We shall not reproduce the intermediate steps because of their tedium. The final results are

$$E[N_{\text{np}}] = \frac{\bar{\alpha} r(1 - \alpha r + \alpha r^2)}{(1 - r)(1 - \alpha r)} \tag{218}$$

as obtained at (195) by a much more attractive method, and

$$\text{Var}[N_{\text{np}}] = \frac{\bar{\alpha}^2 r^2}{(1 - r)^2(1 - \alpha r)^3} [2r(1 - \alpha r^2) - (1 - \alpha r)(1 - \alpha r + \alpha r^2)^2 +$$
$$\frac{\bar{\alpha} r(1 + 2\bar{\alpha} r)(1 - \alpha r + \alpha r^2)}{(1 - r)(1 - \alpha r)} . \tag{219}$$

It is readily confirmed that (218) and (219) reduce to the $M/M/1$ values $r/(1 - r)$ and $r/(1 - r)^2$ when $\bar{\alpha} = 1$.

4.8.3 Numerical values and comments

It is instructive, before making the final attack on W_{np}, to compare the means and standard deviations of N_p and N_{np} on the lines of Table 22. The results of the comparison are given in Table 23.

It is seen that increasing the proportion α of priority messages decreases both the mean number of non-priority messages and its standard deviation. The reverse is the case for priority messages: increasing the priority ratio α increases $E[N_p]$ and the standard deviation. Note that $E[N_p] + E[N_{np}] = r/(1-r)$, the value for the simple $M/M/1$ system.

The congestion induced for non-priority customers under heavy traffic is clearly in evidence. Each of the average number of 90 non-priority messages waiting for $r = 0.99$, $\alpha = 0.9$, could expect an average wait of 909 message transmission times before reaching its destination (see Table 22).

4.8.4 Laplace transform of $g_{np}(t)$ and statistics of W_{np}

We turn finally to the distribution and statistics of W_{np}. It seems improbable that as simple as expression as (191) can be obtained for the probability density function $g_{np}(t)$ (but see the end of this section). However, the convolutions in (179) for $g_{np}(t)$ invite Laplace transformation. We write $\gamma_{np}(z)$ for this Laplace transform. Since it can in principle be made to yield all moments of the random variable W_{np} it is sufficient for practical purposes. Our immediate objective is to verify (196) and to find $\text{Var}[W_{np}]$. A table resembling Table 23 will then be made to examine the fortunes of non-priority messages from the point of view of standard deviation of system time.

In transforming (179) we use the fact that under Laplace transformation $k_n(t)$ and $f_n(t)$ pass respectively into $\kappa^n(z)$ and $\phi^n(z)$, where $\kappa(z)$ and $\phi(z)$ are the Laplace transforms respectively of the probability density functions of the busy period B_1 (see Section 4.4), and the interval X separating successive non-priority service opportunities. $\kappa(z)$ is given by (174), and $\phi(z)$ by (173). Now it happens that for the system under consideration

$$\phi(z) = \kappa(z) . \tag{220}$$

Probabilistically this means that X and B_1 are statistically identical. After a moment's thought this is not surprising, since X is actually the busy period initiated by the non-priority message that has just began transmission. The fact that after the completion of transmission of that message the only messages to be transmitted before the next non-priority service opportunity will be priority messages makes no difference statistically since the service mechanism is unaware of priorities, and the arrival mechanism in this case is entirely that of the priority stream. To confirm (220) analytically it is merely necessary to substitute (174) into (173). We then find

$$\phi(z) = \frac{\mu}{Z - (Z - R_1)^{\frac{1}{2}}} = \frac{2\mu}{Z + R_1} = \frac{Z - R_1}{2\alpha\lambda} = \kappa(z) ,$$

Table 23 — Means and standard deviations of N_p and N_{np}

	N_{np}			N_p	
	Mean	Sdvn		Mean	Sdvn
α	$r = 0.1$		α	$r = 0.1$	
0.1	0.1001	0.3319	0.1	0.0110	0.1055
0.2	0.0891	0.3116	0.2	0.0220	0.1501
0.3	0.0780	0.2901	0.3	0.3312	0.1849
0.4	0.0669	0.2674	0.4	0.0442	0.2147
0.5	0.0558	0.2429	0.5	0.0553	0.2414
0.6	0.0447	0.2163	0.6	0.0664	0.2660
0.7	0.0335	0.1864	0.7	0.0775	0.2889
0.8	0.0224	0.1514	0.8	0.0887	0.3107
0.9	0.0112	0.1065	0.9	0.0999	0.3314
α	$r = 0.5$		α	$r = 0.5$	
0.1	0.9237	1.3502	0.1	0.0763	0.2844
0.2	0.8444	1.2810	0.2	0.1556	0.4181
0.3	0.7618	1.2053	0.3	0.2383	0.5330
0.4	0.6750	1.1212	0.4	0.3250	0.6418
0.5	0.5833	1.0263	0.5	0.4167	0.7500
0.6	0.4857	0.9169	0.6	0.5143	0.8614
0.7	0.3808	0.7876	0.7	0.6192	0.9794
0.8	0.2667	0.6301	0.8	0.7333	1.1075
0.9	0.1409	0.4263	0.9	0.8591	1.2505
α	$r = 0.7$		α	$r = 0.7$	
0.1	2.2106	2.7219	0.1	0.1226	0.3643
0.2	2.0794	2.6462	0.2	0.2539	0.5456
0.3	1.9373	2.5582	0.3	0.3960	0.7103
0.4	1.7811	2.4527	0.4	0.5522	0.8765
0.5	1.6064	2.3214	0.5	0.7269	1.0551
0.6	1.4064	2.1514	0.6	0.9268	1.2572
0.7	1.1708	1.9207	0.7	1.1625	1.4976
0.8	0.8624	1.5896	0.8	1.4509	1.7996
0.9	0.5114	1.0783	0.9	1.8218	2.2039
α	$r = 0.9$		α	$r = 0.9$	
0.1	8.8219	9.4511	0.1	0.1790	0.4436
0.2	8.6224	9.4081	0.2	0.3776	0.6765
0.3	8.3971	9.3534	0.3	0.6029	0.9007
0.4	8.1337	9.2979	0.4	0.8662	1.1442
0.5	7.8136	9.1732	0.5	1.1864	1.4323
0.6	7.4035	9.0061	0.6	1.5965	1.8034
0.7	6.3375	8.7149	0.7	2.1624	2.3301
0.8	5.9657	8.1265	0.8	3.0343	3.1779
0.9	4.3532	6.6176	0.9	4.6468	4.8266
α	$r = 0.95$		α	$r = 0.95$	
0.1	18.8053	19.4737	0.1	0.1947	0.4634
0.2	18.5872	19.4492	0.2	0.4128	0.7098
0.3	18.3363	19.4172	0.3	0.6637	0.9505
0.4	18.0377	19.3724	0.4	0.9623	0.2171
0.5	17.6655	19.3043	0.5	1.3345	0.5417
0.6	17.1707	19.1900	0.6	1.8293	0.7966
0.7	16.4492	18.9713	0.7	2.5508	2.6313
0.8	15.2317	18.4594	0.8	3.7683	3.7879
0.9	12.5433	16.7339	0.9	6.4567	6.4707
α	$r = 0.99$		α	$r = 0.99$	
0.1	98.7922	99.4944	0.1	0.2077	0.4792
0.2	98.5576	99.4890	0.2	0.4424	0.7366
0.3	98.2847	99.4817	0.3	0.7153	0.9909
0.4	97.9549	99.4711	0.4	1.0451	1.2774
0.5	97.5346	99.4544	0.5	1.4654	1.6344
0.6	96.9576	99.4244	0.6	2.0424	2.1297
0.7	96.0722	99.3613	0.7	2.9278	2.9160
0.8	94.4384	99.1883	0.8	4.5616	4.4369
0.9	90.0164	98.3548	0.9	8.9836	8.7516

as stated. There are other reasons why (220) is not surprising, but they are not in place here.

In the following analysis we shall record more detail than usual to help the reader check the steps. Applying the Laplace transformation to (179) according to the rules gives

$$\gamma_{np}(z) = \frac{\mu}{(\mu+z)} [p_0 + \sum_{m \geq 1} a_{m0}\, \kappa^m(z) + \sum_{n \geq 1} b_{0n}\, \kappa^n(z) +$$

$$+ \sum_{m \geq 1} \sum_{n \geq 1} a_{mn}\, \kappa^{n+m}(z) + \sum_{m \geq 1} \sum_{n \geq 1} b_{mn}\, \kappa^{m+n}(z)]$$

$$= \frac{\mu}{\mu+z} [p_0 + \sum_{m \geq 1} a_{m0}\, \kappa^m(z) + B_0(\kappa(z)) +$$

$$+ \sum_{m \geq 1} \kappa^m(z)\, \{A_m(\kappa(z)) - a_{m0}\} + \sum_{m \geq 1} \kappa^m(z)\, B_m(\kappa(z))]$$

$$= \frac{\mu}{\mu+z} [p_0 + \sum_{m \geq 0} B_m(\kappa(z))\, \kappa^m(z) + \sum_{m \geq 1} A_m(\kappa(z))\, \kappa^m(z)].$$

Hereafter we shall usually omit specific reference to the argument z of $\kappa(z)$. Substituting (202) and (210) gives the considerably reduced expression

$$\gamma_{np}(z) = \frac{\mu}{(\mu+z)(1 - \kappa\eta(\kappa))} \left[p_0 + \frac{B_0(\kappa)\, \eta(\kappa)}{\sigma(\kappa)} \right]. \qquad (221)$$

Substituting (211) we then have

$$\gamma_{np}(z) = \frac{\mu\, p_0}{(\mu+z)(1 - \kappa\eta(\kappa))} \left[1 + \frac{\kappa\eta(\kappa)\{\eta(\kappa) - r(1 - \bar{\alpha}\kappa)\}}{\alpha r \kappa - \eta(\kappa)} \right]$$

$$= \frac{\mu\, p_0[\kappa\{\eta^2(\kappa) - \eta(\kappa)\, X(\kappa) + \alpha r\} - \eta(\kappa)(1 - \kappa)}{(\mu+z)(1 - \kappa\eta(\kappa))(\alpha r \kappa - \eta(\kappa))}.$$

Since η satisfies (205) this reduces to

$$\gamma_{np}(z) = \frac{\mu p_0 \eta(\kappa)(1 - \kappa(z))}{(\mu+z)(1 - \kappa(z)\, \eta(\kappa))\, (\eta(\kappa) - \alpha r \kappa(z))}. \qquad (222)$$

Naturally we should check this by putting $z=0$ which should give $\gamma_{np}(0)=1$. However, there appears again the difficulty encountered already in the form (211) that both numerator and denominator vanish when $z=0$. This can be mitigated by using the exponential substitution introduced in (215) where in this case $x = \kappa(z)$. After some algebra we arrive at the considerably simplified form

$$\gamma_{np}(z) = \frac{p_0\,\mu\,\bar{\alpha}}{\alpha(\mu+z)\left\{\left(\frac{r}{\alpha}\right)^{\frac{1}{2}} - e^{-u}\right\}\left\{e^u - \left(\frac{r}{\alpha}\right)^{\frac{1}{2}}\right\}}, \qquad (223)$$

in which the relation between u, z, and κ is given by

$$2(\alpha r)^{\frac{1}{2}} \cosh u = 1 + r - \bar{\alpha} r\,\kappa(z). \qquad (224)$$

After checking that $\gamma_{np}(0) = 1$ (recall that when $z \to 0$, $\kappa(z) \to 1$ ($\alpha\lambda < \mu$), $u \to u_1$ where $\exp(-u_1) = (\alpha r)^{\frac{1}{2}}$) we write

$$\ln \gamma_{np}(z) = \ln\left(\mu\,p_0\,\frac{\bar{\alpha}}{\alpha}\right) + u - \ln(\mu+z) - \ln M - \ln N \qquad (225)$$

where

$$M = \left(\frac{r}{\alpha}\right)^{\frac{1}{2}} e^u - 1,$$

$$N = e^u - \left(\frac{r}{\alpha}\right)^{\frac{1}{2}}. \qquad (226)$$

Then

$$\gamma'_{np}(z)/\gamma_{np}(z) = \frac{du}{dz} - \frac{1}{(\mu+z)} - \left(\frac{M_u}{M} + \frac{N_u}{N}\right)\frac{du}{dz}. \qquad (227)$$

Noting that

$$\frac{du}{dz} = \frac{-\bar{\alpha} r}{(\alpha r)^{\frac{1}{2}}(e^u - e^{-u})}\frac{d\kappa}{dz},$$

$$\frac{d\kappa}{dz} = \frac{-\kappa(z)}{R_1(z)}$$

Sec. 4.8] Non-priority system state and the form of $g_{np}(t)$

$$M_u = \left(\frac{r}{\alpha}\right)^{\frac{1}{2}} e^u,$$

$$N_u = e^u,$$

we get

$$\left(\frac{du}{dz}\right)_{z=0} = \frac{\bar{\alpha}r}{\mu(1-\alpha r)^2},$$

$$\left(\frac{M_u}{M}\right)_{z=0} = \frac{1}{\bar{\alpha}},$$

$$\left(\frac{N_u}{N}\right)_{z=0} = \frac{1}{1-r}.$$

Inserting these in (227) we obtain

$$E[W_{np}] = -\gamma'_{np}(0) = \frac{1-\alpha r + \alpha r^2}{\mu(1-r)(1-\alpha r)} = \frac{E[N_{np}]}{\bar{\alpha}\lambda} \tag{228}$$

in agreement with (195) and (196).

To get the variance we differentiate (227) again, obtaining

$$\frac{\gamma''_{np}(z)}{\gamma_{np}(z)} - \left[\frac{\gamma'_{np}(z)}{\gamma_{np}(z)}\right]^2 = \frac{d^2u}{dz^2} + \frac{1}{(\mu+z)^2} - \left(\frac{M_u}{M} + \frac{N_u}{N}\right)\frac{d^2u}{dz^2} -$$

$$- \left(\frac{du}{dz}\right)^2 \left[\frac{M_{uu}}{M} - \left(\frac{M_u}{M}\right)^2 + \frac{N_{uu}}{N} - \left(\frac{N_u}{N}\right)^2\right]. \tag{229}$$

Then

$$\text{Var}[W_{np}] = \frac{1}{\mu^2} + \left(1 - \frac{1}{\bar{\alpha}} - \frac{1}{\bar{r}}\right)\left[\frac{d^2u}{dz^2}\right]_{z=0} - \frac{\bar{\alpha}^2 r^2}{\mu^2(1-\alpha r)^4}\left[\frac{1}{\bar{\alpha}}\left(1 - \frac{1}{\bar{\alpha}}\right) + \frac{1}{\bar{r}}\left(1 - \frac{1}{\bar{r}}\right)\right]$$

where $\bar{r} = 1 - r$, and since $M_{uu} = M_u$, $N_{uu} = N_u$.

Now

$$\frac{d^2u}{dz^2} = \frac{-(\bar{\alpha}r)}{\mu^2(1-\alpha r)^5}(2 - 2\alpha r + \bar{\alpha}r + \alpha\bar{\alpha}r^2),$$

and

$$1 - \frac{1}{\bar{\alpha}} - \frac{1}{\bar{r}} = \frac{(\bar{\alpha}-1)(\bar{r}-1)-1}{\bar{\alpha}\bar{r}} = \frac{-(1-\alpha r)}{\bar{\alpha}\bar{r}}.$$

Consequently

$$\text{Var}[W_{np}] = \frac{1}{\mu^2} + \frac{r(2-2\alpha r + \bar{\alpha} r + \alpha\bar{\alpha} r^2)}{\mu^2(1-\alpha r)^4(1-r)} + \frac{\bar{\alpha}^2 r^2}{\mu^2(1-\alpha r)^4}\left(\frac{\alpha}{\bar{\alpha}^2} + \frac{r}{\bar{r}^2}\right). \tag{230}$$

A check on this can be provided by putting $\bar{\alpha} = 1$. All traffic is then homogeneous and $\text{Var}[W_{np}]$ reduces, as it should, to $(\mu(1-r))^{-2}$. It is depressing for the non-priority message that the standard deviation in this case should appear to behave like the inverse square of $1 - \alpha r$ as well as the inverse of $(1-r)^{1/2}$.

Substitution of (224) back into (223) gives the even simpler form

$$\gamma_{np}(z) = \frac{\mu p_0}{(\mu + z)(1 - r\kappa(z))},$$

which yields the moments much more easily than above. Moreover, expansion of the denominator in a series of ascending powers of $\kappa(z)$ and inversion by term leads to an expression for the distribution function

$$G_{np}(t) = \int_0^t g_{np}(v) \, dv$$

of W_{np} where t is measured in units of μ^{-1}. This is

$$G_{np}(t) = rp_0 \int_0^t \{1 - e^{-(t-v)}\} e^{-v(1+\alpha r)} \sum_{n \geq 0} \left(\frac{r}{\alpha}\right)^{n/2} \{I_n(2v\sqrt{(\alpha r)}) - I_{n+2}(2v\sqrt{(\alpha r)})\} \, dv$$

entailing modified Bessel functions. In spite of its formidable appearance this expression is not hard to calculate numerically.

4.8.5 Numerical values and comments

Table 24 complements Table 22 by giving numerical values of the standard deviations

Table 24 — Standard deviation of equilibrium system times W_p and W_{np} of priority and non-priority streams as functions of traffic intensity r and priority ratio α

α	Priority $r = 0.1$	Non-priority	α	Priority $r = 0.5$	Non-priority
0.1	1.0926	1.1143	0.1	1.3532	2.1073
0.2	1.0945	1.1176	0.2	1.3878	2.2312
0.3	1.0963	1.1210	0.3	1.4276	2.3750
0.4	1.0983	1.1245	0.4	1.4737	2.5434
0.5	1.1002	1.1282	0.5	1.5275	2.7420
0.6	1.1023	1.1320	0.6	1.5908	2.9785
0.7	1.1043	1.1360	0.7	1.6659	3.2634
0.8	1.1065	1.1402	0.8	1.7559	3.6107
0.9	1.1088	1.1445	0.9	1.8653	4.0406

α	Priority $r = 0.7$	Non-priority	α	Priority $r = 0.9$	Non-priority
0.1	1.4325	3.6189	0.1	1.4817	11.0770
0.2	1.4935	3.9621	0.2	1.5724	12.4141
0.3	1.5678	4.3805	0.3	1.6905	14.1163
0.4	1.6599	4.8995	0.4	1.8485	16.3529
0.5	1.7759	5.5569	0.5	2.0671	19.4154
0.6	1.9249	6.4111	0.6	2.3830	23.8479
0.7	2.1210	7.5573	0.7	2.8691	30.7901
0.8	2.3876	9.1600	0.8	3.6916	43.0697
0.9	2.7654	11.5282	0.9	5.3314	69.7786

α	Priority $r = 0.95$	Non-priority	α	Priority $r = 0.99$	Non-priority
0.1	1.4893	22.2044	0.1	1.4939	111.1074
0.2	1.5876	24.9545	0.2	1.5983	124.9904
0.3	1.7179	28.4799	0.3	1.7387	142.8376
0.4	1.8960	33.1598	0.4	1.9341	166.6288
0.5	2.1492	39.6656	0.5	2.2183	199.9248
0.6	2.5288	49.3087	0.6	2.6582	249.8373
0.7	3.1446	65.0270	0.7	3.4072	332.9182
0.8	4.2799	94.9590	0.8	4.9104	498.5409
0.9	6.9601	172.1501	0.9	9.2282	988.8497

Note: time unit is mean processing time.

$(\mathrm{Var}[W_p])^{\frac{1}{2}}$, $(\mathrm{Var}[W_{np}])^{\frac{1}{2}}$ calculated from (193) and (230) respectively. After the discussion that has preceded, it will be no surprise to observe the adverse effect of this priority discipline on the system time standard deviation of non-priority messages at heavy traffic levels. If there were no priority discipline, and customers were treated as homogeneous units with first come, first served discipline, the system times W would be exponentially distributed with parameter $\mu(1-r)$ and $E[W] = \mathrm{SDVN}(W) = (\mu(1-r))^{-1}$. We can now see that the effect of priority for the fortunate customers is to confer on them a system time distribution that is 'better' than exponential, and much better at practical traffic levels. The fate of the less fortunate non-priority customers is a W distribution that is worse — much worse — than exponential as $r \to 1$ and $\alpha \to 1$.

4.9 MANAGEMENT OF A PRIORITY SYSTEM

4.9.1 Basic considerations

A simple model has exposed the principles with which management of a message system with priority facility has to be familiar. Without the knowledge of the features and requirements of particular systems the theoretician, on the other hand, has to abstain from sweeping pronouncements. Detailed advice can be given only in a rational manner after examination of the problems in the field, followed probably by further detailed analysis and consultation. Here we can only underline what is common to most demand and supply situations.

In the command and control context under operational conditions the demand to transmit messages is likely to be high in relation to the number of communication channels available, and the system will therefore be operating most of the time under heavy traffic conditions unless measures are taken. That is to say, the parameter r introduced in the text will tend to have a value in the neighbourhood of unity. In addition to their natural urge to pass on information, originators of messages are likely in many instances to possess an exaggerated notion of the importance of the content. This, coupled with the belief that it will arrive at the destination more quickly, encourages a possibly unjustified allocation of priority to the message. This tends to increase the ratio α of priority messages in the total message stream. Originators of messages who would normally avoid using priority, disturbed by the long delays that experience shows their diffidence produces, are then likely, in sheer self defence, themselves to request priority, thus making matters worse.

There are accordingly two major considerations:

(i) management of traffic intensity;
(ii) at a given level of traffic intensity, management of the priority mechanism.

4.9.2 Management of the traffic intensity

In the text, traffic intensity has been denoted by the parameter r, whose numerical value, expressed in terms of a general service system, is the average number of demands for service made during the mean service time of a customer. r has also the convenient interpretation when less than unity as the probability that the service is busy. When $r \geq 1$ the service will virtually always be busy and the system is out of hand. In this case the system is said to be in a state of congestion, and the number of customers waiting, or turned away, will tend to increase with the passage of time. The waiting time for service will also increase without limit.

The first management principle is accordingly to maintain r at a numerical value below unity. There are only two ways of doing this: adjustment of the rate of providing service, and management of the demand. In some contexts it is possible to cause the service to respond to increasing demand by working faster, but in the communication context this does not seem obviously a practical proposition. If technology permits it, it should not be overlooked as a possibilty.

Sec. 4.9]   **Management of a priority system**   113

A simpler method is to provide more service channels which, in the communication context, means more circuits. A detailed analysis of a two-channel system with priority, on the lines of what has been done in the first part of this paper, would be illuminating, and has not, to this writer's knowledge, ever been attempted in this detail.†. Qualitatively we can infer what the effect might be. To add a second channel identical in service characteristics with the first to a non-priority system where traffic intensity is r reduces the probability that both channels are busy (i.e. that the service is entirely occupied) from r to $r^2/(2+r)$. For example, if $r = 0.99$, $r^2/(2+r)$ is roughly 0.33, and the effect is to transform a system under heavy traffic to one with light traffic. Moreover, the mean waiting time is reduced from $(1-r)^{-1}$ to $(1-r^2/4)^{-1}$ in units of mean service time of a customer. Again, with $r = 0.99$, this means a reduction from 100 units to 1.325. We have seen that the effect of assigning priority to a proportion of customers is to reduce their mean waiting time relative to a non-priority system, and to increase that of non-priority customers. But, at light traffic levels there is little difference between the system times of either species of customer, so priority becomes less important as a mere factor in getting one's message to its destination promptly, and the psychological pressure to assign priority is reduced.

Restriction of demand is the second method available for the management of traffic. It is a common assumption that demand and service are independent, but, in real life, this is not true. It is common experience that to upgrade a computer system has the initial effect of reducing the load on a new system, but that as customers become aware of the new facilities available, demand appears where there was none before, and the new system before long again becomes saturated. The lesson is that a very rigorous discipline must be imposed on demand, all the more in an operational context where the arrival of a message may be a matter of life and death. This may be done by the enforcement of strict protocols, the effect of which is to filter out and reject selectively a proportion of the would-be customers. Alternatively, and easier to apply, but potentially more dangerous in the sense of risking the rejection of important messages, a limit may be imposed on the number of messages permitted to be resident in the system. If this were 50, say, the numbers shown in Table 23 for $r = 0.99$ would not arise. Such systems are called 'loss-systems' and were the subject of study by Erlang, the father of the subject, in the early twentieth century. Again, the detailed effect of such a measure in the context of a priority system has not, to this writer's knowledge, been evaluated.

4.9.3 Management of priority mechanism

We have studied a particular mechanism described as 'head-of-the-line priority' in the literature. Whether this applies or not, a rational procedure for the allocation of priority must take into account the loss in value of the commodity due to its late arrival, as well as the cost of implementing the system. Here we shall ignore the latter. The former might be evaluated by supposing that the value of the information transmitted decreases exponentially. That is to say we might suppose that the decrease is at a constant fraction z of current value per unit time late. It is as though

---

† Subsequently to writing this a detailed analysis has been made. The results are provisionally available in [15] and [16].

interest is compounded negatively and continuously on the initial value. If the time late W has probability density function $g(t)$, the expected value of the information on arrival, given that the initial value was unity, is given by

$$E[e^{-zW}] = \int_0^\infty e^{-zt} g(t) \, dt = \gamma(z) ,$$

which will be recognized to be the Laplace transform of $g(t)$. The theory in the first part of the chapter provides an expression for $\gamma_{np}(z)$ in (223). $\gamma_p(z)$ can be obtained from (191): specifically

$$\gamma_p(z) = \frac{(1 - \alpha r) \mu}{\alpha(z + \mu - \alpha r \mu)} - \frac{\mu \bar{\alpha}}{\alpha(\mu + z)} .$$

Finally, from (192) for a system without priority we have

$$\gamma(z) = \frac{\mu(1 - r)}{z + \mu - \mu r} .$$

4.9.4 Numerical values and comments

Table 25 gives values of $\gamma(z), \gamma_p(z), \gamma_{np}(z)$ and the weighted sum $\alpha \gamma_p(z) + \bar{\alpha} \gamma_{np}(z)$, as functions of the priority ratio α for the values of r used in Tables 22 to 24, spanning light, medium, and heavy traffic. The unit of time is fixed by putting $\mu = 1$ as was done in Tables 22 and 24. The values are repeated for $z = 0.1$ and $z = 0.5$. These are, as it were, interest rates. It will be seen that the second column $\gamma(z)$ is constant. It does not depend upon α, and it tells us how much the value of a message can be expected to deteriorate under a single-channel, non-priority system. The next column $\gamma_p(z)$, shows how much the values of priority messages decrease on average, and it is seen that, as they arrive earlier, their value has decreased less, as would be expected. The next column contains $\gamma_{np}(z)$ and shows that non-priority messages decrease in value on average more than if there were no priority system in operation.

The weighted combination whose value is given in the next column shows the value of any message that enters the system at priority ratio value α. It is seen that the value rises to a maximum as α increases and then decreases again. This means that a value of α exists for this kind of priority system such that the devaluation of any message entering the system is least. A similar finding is likely to hold for other priority systems.

Although the criterion chosen for the evaluation of the consequence of delaying the arrival of information may not be the most appropriate, the argument shows that an optimal priority ratio may well exist for each level of traffic intensity r (in fact there seems to be little variation) which is such that the value of messages, priority and otherwise, suffers the least decrease. This is also a topic deserving a deeper analysis.

Table 25 — Reduced value of message due to late arrival

α	$\gamma(z)$	$z = 0.1$, $r = 0.1$ $\gamma_p(z)$	$\gamma_{np}(z)$	$\alpha\gamma_p(z) + \bar{\alpha}\gamma_{np}(z)$	α	$\gamma(z)$	$z = 0.1$, $r = 0.5$ $\gamma_p(z)$	$\gamma_{np}(z)$	$\alpha\gamma_p + \bar{\alpha}\gamma_{np}$
0.1	0.9	0.9007	0.8999	0.9000	0.1	0.8333	0.8658	0.8304	0.8339
0.2	0.9	0.9007	0.8998	0.9000	0.2	0.8333	0.8636	0.8272	0.8345
0.3	0.9	0.9006	0.8998	0.9000	0.3	0.8333	0.8612	0.8238	0.8350
0.4	0.9	0.9005	0.8997	0.9000	0.4	0.8333	0.8586	0.8201	0.8355
0.5	0.9	0.9004	0.8996	0.9000	0.5	0.8333	0.8556	0.8162	0.8359
0.6	0.9	0.9003	0.8995	0.9000	0.6	0.8333	0.8522	0.8119	0.8361
0.7	0.9	0.9003	0.8994	0.9000	0.7	0.8333	0.8485	0.8074	0.8362
0.8	0.9	0.9002	0.8994	0.9000	0.8	0.8333	0.8442	0.8025	0.8358
0.9	0.9	0.9001	0.8993	0.9000	0.9	0.8333	0.8392	0.7973	0.8350
		$z = 0.1$, $r = 0.9$					$z = 0.1$, $r = 0.95$		
0.1	0.5	0.8281	0.4827	0.5172	0.1	0.3333	0.8232	0.3164	0.3671
0.2	0.5	0.8202	0.4642	0.5354	0.2	0.3333	0.8142	0.2988	0.4019
0.3	0.5	0.8105	0.4445	0.5543	0.3	0.3333	0.8031	0.2807	0.4374
0.4	0.5	0.7985	0.4237	0.5737	0.4	0.3333	0.7891	0.2620	0.4729
0.5	0.5	0.7832	0.4018	0.5925	0.5	0.3333	0.7709	0.2431	0.5070
0.6	0.5	0.7630	0.3791	0.6094	0.6	0.3333	0.7461	0.2241	0.5373
0.7	0.5	0.7350	0.3558	0.6213	0.7	0.3333	0.7105	0.2055	0.5590
0.8	0.5	0.6988	0.3324	0.6215	0.8	0.3333	0.6551	0.1875	0.5616
0.9	0.5	0.6270	0.3094	0.5952	0.9	0.3333	0.5566	0.1708	0.5180
		$z = 0.1$, $r = 0.99$							
0.1		0.8192	0.0909		0.1	0.1577			0.0842
0.2		0.8093	0.0909		0.2	0.2239			0.0776
0.3		0.7970	0.0909		0.3	0.2888			0.0710
0.4		0.7812	0.0909		0.4	0.3512			0.0645
0.5		0.7603	0.0909		0.5	0.4093			0.0583
0.6		0.7312	0.0909		0.6	0.4596			0.0523
0.7		0.6880	0.0909		0.7	0.4956			0.0467
0.8		0.6169	0.0909		0.8	0.5018			0.0416
0.9		0.4785	0.0909		0.9	0.4343			0.0371

$z = 0.5$, $r = 0.1$

α	γ(z)	γ_p(z)	γ_mp(z)	αγ_p + ᾱγ_mp
0.1	0.6429	0.6443	0.6427	0.6429
0.2	0.6429	0.6441	0.6426	0.6429
0.3	0.6429	0.6439	0.6425	0.6430
0.4	0.6249	0.6438	0.6424	0.6430
0.5	0.6249	0.6437	0.6423	0.6430
0.6	0.6429	0.6435	0.6422	0.96430
0.7	0.6429	0.6434	0.6421	0.6430
0.8	0.6429	0.6432	0.6420	0.6430
0.9	0.6429	0.6430	0.6419	0.6429

$z = 0.5$, $r = 0.9$

α	γ(z)	γ_p(z)	γ_mp(z)	αγ_p(z) + ᾱγ_mp(z)
0.1	0.1666	0.4539	0.1617	0.1909
0.2	0.1666	0.4393	0.1568	0.2133
0.3	0.1666	0.4227	0.1522	0.2333
0.4	0.1666	0.4035	0.1476	0.2500
0.5	0.1666	0.3819	0.1433	0.2621
0.6	0.1666	0.3541	0.1391	0.2681
0.7	0.1666	0.3218	0.1352	0.2658
0.8	0.1666	0.2820	0.1314	0.2519
0.9	0.9	0.2318	0.1279	0.2214

$z = 0.1$, $r = 0.5$

α	γ(ζ)	γ_p(z)	γ_mp(z)	αγ_p + ᾱγ_mp
0.1	0.5	0.5517	0.4972	0.5027
0.2	0.5	0.5476	0.4940	0.5050
0.3	0.5	0.5432	0.4915	0.5070
0.4	0.5	0.5385	0.4887	0.5086
0.5	0.5	0.5333	0.4858	0.5096
0.6	0.5	0.5278	0.4829	0.5098
0.7	0.5	0.5217	0.4800	0.5092
0.8	0.5	0.5151	0.4771	0.5075
0.9	0.5	0.5079	0.4743	0.5046

$z = 0.5$, $r = 0.95$

α	r(z)	γ_p(z)	γ_mp(z)	αγ_p + ᾱγ_mp
0.1	0.0909	0.4413	0.0876	0.1231
0.2	0.0909	0.4249	0.0845	0.1526
0.3	0.0909	0.4060	0.0815	0.1788
0.4	0.0909	0.3839	0.0786	0.2007
0.5	0.0909	0.3577	0.0759	0.2168
0.6	0.0909	0.3262	0.0732	0.2250
0.7	0.0909	0.2874	0.0708	0.2224
0.8	0.0909	0.2387	0.0686	0.2047
0.9	0.0909	0.1757	0.0664	0.1648

$z = 0.5$, $r = 0.99$

α				
0.1	0.0188	0.4311	0.0196	
0.2	0.0180	0.4132	0.0196	
0.3	0.0173	0.3923	0.0196	
0.4	0.0166	0.3677	0.0196	0.0600
0.5	0.0159	0.3383	0.0196	0.0971
0.6	0.0152	0.3025	0.0196	0.1298
0.7	0.0147	0.2577	0.0196	0.1570
0.8	0.0141	0.2005	0.0196	0.1771
0.9	0.0136	0.1248	0.0196	0.1876
				0.1348
				0.1633
				0.1137

Notes:

- z is the 'compound interest' rate at which the value decreases.
- r is the traffic intensity.
- α is the proportion of traffic assigned priority.
- $\gamma(z)$ is the average message value under a non-priority system.
- $\gamma_p(z)$ is the average message value under a priority system.
- $\gamma_{mp}(z)$ is the average message value under a non-priority system.
- $\alpha\gamma_p + \bar{\alpha}\gamma_{mp}$ is the weighted average.

In conclusion we repeat that traffic intensity r must be controlled first. When this has been done, an optimal value should be sought upon which to base the selection of priority messages in such a way as to minimize the damage inflicted by delays.

5
Single-channel service with alternating priority

5.1 INTRODUCTION TO ALTERNATING PRIORITY

The following material is complementary to the analysis of a head-of-the-line priority message system reported in Chapter 4. The so-called 'alternating priority' discipline is investigated. This is such that the service devotes itself entirely to a single priority class until no more of this class is left. As before, the model envisages two classes only, namely, a proportion α of the total traffic which we call the P-class, and the remaining proportion $\bar{\alpha} = 1 - \alpha$ constitutes the NP-class. The question of priority in this case is really non-existent. It arises only at the beginning of operations when there may be a rule which says 'choose P-customers first'. After that the service switches from one class to the other, or to the first arrival in the eventuality that the service becomes completely idle on completing the service of one type. This is the discipline encountered sometimes at road intersections where traffic in one direction has precedence until none is left. It is of special interest in being a self-regulating mechanism and easy to implement. For earlier analysis of this system see [14] and references therein.

The evolution of the system is a series of service periods each composed of a sum of busy periods. If this system has an advantage in comparison with head-of-the-line priority it is that it avoids the build-up of exceptionally long queues and waiting times of either traffic type under all but the heaviest traffic conditions when they are unavoidable. P customers are on the whole worse off than under head-of-the-line priority, but the other class is better favoured. In the context of messages, their information content will reach the destination in a more balanced manner.

5.2 STATISTICAL DESCRIPTION OF SYSTEM STATE

We begin, as before, by investigating the state probabilities on the assumption of statistical equilibrium. The underlying processes are X and Y, the numbers in the P

Statistical description of system state

and NP classes, respectively, and Z, a random variable equal to 1 when P messages are being processed, and equal to 0 when NP messages are being processed. Thus

$$p_0 = P[X=0, Y=0] = 1-r \qquad (231)$$

for the single channel system with Poisson arrivals, mean rate λ, and exponential service time with mean μ^{-1}. As usual $r = \lambda/\mu$. In this case Z does not enter: otherwise we write

$$\begin{aligned} a_{mn} &= P[X=m, \ Y=n, \ Z=1] & (m \geq 1, \ n \geq 0), \\ b_{mn} &= P[X=m, \ Y=n, \ Z=0] & (m \geq 0, \ n \geq 1). \end{aligned} \qquad (232)$$

There are four sets of equations to be solved. These show what a moment's thought will confirm, namely that a_{mn} and b_{mn} are symmetrical with respect to α and $\bar{\alpha}$, and m and n. The equations are:

$$(1+r)a_{10} = a_{20} + b_{11} + \alpha r p_0$$

....

$$(1+r)a_{1n} = a_{2n} + \bar{\alpha} r a_{1\ n-1} \qquad (n \geq 1) \qquad (233)$$

$$(1+r)a_{m0} = a_{m+1\ 0} + b_{m1} + \alpha r a_{m-1\ 0} \qquad (m \geq 2)$$

....

$$(1+r)a_{mn} = a_{m+1\ n} + \alpha r a_{m-1\ n} + \bar{\alpha} r a_{m\ n-1} \qquad (n \geq 1) \qquad (234)$$

$$(1+r)b_{01} = b_{02} + a_{11} + \bar{\alpha} r p_0$$

....

$$(1+r)b_{m1} = b_{m2} + \alpha r b_{m-1\ 1} \qquad (m \geq 1) \qquad (235)$$

$$(1+r)b_{0n} = b_{0\ n+1} + a_{1n} + \bar{\alpha} r b_{0\ n-1} \qquad (n \geq 2)$$

....

$$(1+r)b_{mn} = b_{m\ n+1} + \alpha r b_{m-1\ n} + \bar{\alpha} r b_{m\ n-1} \qquad (m \geq 1). \qquad (236)$$

An ignorant onlooker will merely observe the total number present in the system. Thus

$$a_{mn} + b_{mn} = p_{m+n} = (1-r)r^{m+n}, \tag{237}$$

a standard result for the $M/M/1$ system.

It is also convenient sometimes to rewrite (233) to (236) in such a way as to emphasize the constancy of the total number present. Thus we have

$$\sum_{m=1}^{n} a_{m\,n-m} + \sum_{m=0}^{n-1} b_{m\,n-m} = p_n = (1-r)r^n, \quad (n \geq 1) \tag{238}$$

and

$$(1+r)a_{n0} = a_{n+1\,0} + b_{n1} + \alpha r a_{n-1\,0}$$
$$(1+r)a_{n-1\,1} = a_{n1} + \alpha r a_{n-2\,1} + \bar{\alpha} r a_{n-1\,0}$$
$$(1+r)a_{n-2\,2} = a_{n-1\,2} + \alpha r a_{n-3\,2} + \bar{\alpha} r a_{n-2\,1}$$

$$\ldots\ldots\ldots\ldots\ldots\ldots\ldots\ldots\ldots\ldots$$

$$(1+r)a_{1\,n-1} = a_{2\,n-1} + \bar{\alpha} r a_{1\,n-2}$$
$$(1+r)b_{n-1\,1} = b_{n-1\,2} + \alpha r b_{n-2\,1}$$
$$(1+r)b_{n-2\,2} = b_{n-2\,3} + \alpha r b_{n-3\,2} + \bar{\alpha} r b_{n-2\,1}$$

$$\ldots\ldots\ldots\ldots\ldots\ldots\ldots\ldots\ldots\ldots$$

$$(1+r)b_{0n} = a_{1n} + b_{0\,n+1} + \bar{\alpha} r b_{0\,n-1} \tag{239}$$

Addition of these equations and use of (238) gives

$$(1+r)p_n = p_{n+1} + rp_{n-1}, \quad (n \geq 1) \tag{240}$$

the familiar state equations for $M/M/1$ with solution

$$p_n = (1-r)r^n, \quad (n \geq 0). \tag{241}$$

5.3 A GENERATING FUNCTION APPROACH TO THE STATE PROBABILITIES

There are various ways of attempting to find the state probabilities. One which we used in the head-of-the-line analysis is to attempt to solve the basic difference equations directly. But, perversely, in spite of the symmetry, this approach seems

Sec. 5.3] A generating function approach to the state probabilities

fraught with difficulties owing to awkward boundary conditions. An alternative is to use a generating function technique. For this purpose we introduce

$$A_m(x) = \sum_{n \geq 0} a_{mn} x^n \qquad (m \geq 1, |x| \leq 1) \tag{242}$$

$$B_n(x) = \sum_{m \geq 0} b_{mn} x^m \qquad (n \geq 1, |x| \leq 1).$$

From (233) we get

$$(1+r) A_1(x) = A_2(x) + b_{11} + rp_0 + \bar{\alpha}rx A_1(x), \tag{243}$$

and from (234) for $m \geq 2$,

$$(1+r) A_m(x) = A_{m+1}(x) + b_{m1} + \alpha r A_{m-1}(x) + \bar{\alpha}rx A_m(x). \tag{244}$$

These can be rearranged to give:

$$(1+r-\bar{\alpha}rx) A_1(x) = A_2(x) + b_{11} + \alpha r p_0$$
$$(1+r-\bar{\alpha}rx) A_m(x) = A_{m+1}(x) + b_{m1} + \alpha r A_{m-1}(x) \qquad (m \geq 2). \tag{245}$$

By symmetry these have companions

$$(1+r-\alpha rx) B_1(x) = B_2(x) + a_{11} + \bar{\alpha} r p_0$$
$$(1+r-\alpha rx) B_n(x) = B_{n+1}(x) + a_{1n} + \bar{\alpha}r B_{n-1}(x) \qquad (n \geq 2). \tag{246}$$

Equations (245) and (246) can be combined if we next introduce the double generating functions:

$$A(x,y) = \sum_{m \geq 1} A_m(x) y^m, \tag{247}$$

$$B(x,y) = \sum_{n \geq 1} B_n(x) y^n, \qquad |y| \leq 1.$$

We then obtain

$$(1+r-\bar{\alpha}x)A(x,y) = (A(x,y) - yA_1(x))/y + (B_1(y) - b_{01}) + \alpha r p_0 y + \alpha r y A(x,y) \quad (248)$$

giving

$$A(x,y) = \frac{y(A_1(x) - B_1(y) + b_{01} - \alpha r p_0 y)}{\alpha r y^2 - y(1+r-\bar{\alpha}rx) + 1} . \quad (249)$$

By symmetry we have

$$B(x,y) = \frac{y(B_1(x) - A_1(y) + a_{10} - \bar{\alpha} r p_0 y)}{\alpha r y^2 - y(1+r-\alpha rx) + 1} . \quad (250)$$

Equations (249) and (250) in principle contain all information about the system state. This statement can be rendered more precise by noting that $A(x,y)$ and $B(x,y)$ are finite for $|x| \leq 1$ and $|y| \leq 1$. Moreover, the following relation must hold:

$$p_0 + A(1,1) + B(1,1) = 1 . \quad (251)$$

This does not appear to be immediately helpful since the denominators of (249) and (250) both vanish when $x = y = 1$, and thus

$$A_1(1) - B_1(1) = \alpha r p_0 - b_{01} = a_{10} - \bar{\alpha} r p_0 , \quad (252)$$

or

$$a_{10} + b_{01} = r p_0 = p_1 ,$$

which is (238) with $m = n = 1$.

Let

$$X = 1 + r - \bar{\alpha} r x ,$$
$$R = (x^2 - 4\alpha r)^{\frac{1}{2}} , \quad (253)$$

and let $\underline{X}$, $\underline{R}$ denote the corresponding quantities with α and $\bar{\alpha}$ interchanged. Similarly let

$$Y(x) = (X - R)/2\alpha r ,$$
$$\underline{Y}(x) = (\underline{X} - \underline{R})/2\bar{\alpha}r . \quad (254)$$

Sec. 5.3]   A generating function approach to the state probabilities

The denominators of (249) and (250) can then be written respectively as

$$\alpha r(y - Y)(y - 1/\alpha r Y)$$
$$\bar{\alpha} r(y - \underline{Y})(y - 1/\bar{\alpha} r \underline{Y}) .$$

Since

$$(1/\alpha r Y - 1)(1 - Y) = 1/\alpha r Y - 1/\alpha r - 1 + Y = \frac{1 + r - \bar{\alpha} r x}{\alpha r} - \frac{1}{\alpha r} - 1 = \frac{\bar{\alpha} r (1 - x)}{\alpha},$$

it is clear that when $x < 1$, $Y < 1$. Moreover, provided that $\alpha r < 1$, $Y \to 1$ as $X \to 1$. We can accordingly write

$$A(x,y) = \frac{y(B_1(y) - A_1(x) + \alpha r p_0 y - b_{01})}{\alpha r(y - Y)(1/\alpha r Y - y)} \tag{255}$$

and, *mutatis mutandis*,

$$B(x,y) = \frac{y(A_1(y) - B_1(x) + \bar{\alpha} r p_0 y - a_{10})}{\bar{\alpha} r(y - \underline{Y})(1/\bar{\alpha} r \underline{Y} - y)} . \tag{256}$$

Since $A(1,1)$ and $B(1,1)$ are finite the numerators of A and B must vanish when $x \leq 1$ respectively when $y = Y, \underline{Y}$. Thus we conclude that

$$A_1(x) = B_1(Y(x)) - b_{01} + \alpha r Y(x) p_0 \tag{257}$$
$$B_1(x) = A_1(\underline{Y}(x)) - a_{10} + \bar{\alpha} r \underline{Y}(x) p_0 . \tag{258}$$

$A_1(x)$ can be interpreted as the generating function of the number of non-priority messages waiting just before the end of a priority class service period, for the probability that such a service period terminates in any small time interval $(t, t + h)$ is $\mu h + \sum_{n \geq 0} a_{1n} + o(h)$. Apart from a constant $A_1(x)$ serves the same purpose for the epoch just after the end of a priority class service period. A similar interpretation holds for $B_1(x)$ with respect to the number of priority messages waiting at the end of a non-priority class service period.

It is also worth pointing out that $Y(x)$ and $\underline{Y}(x)$ have a concrete interpretation in terms of the system dynamics. $Y(x)$ is the generating function of the number of NP arrivals during the busy period initiated by the service of a single P message. For, from M/M/1 theory, that busy period has probability density function $k(t)$ with Laplace transform

$$\kappa(z) = \frac{z + \mu + \alpha\lambda - ((z + \mu + \alpha\lambda)^2 - 4\alpha\lambda\mu)^{\frac{1}{2}}}{2\alpha\lambda},$$

while the said generating function is

$$\int_0^\infty \exp(-\bar{\alpha}\lambda(1-x)t)\, k(t)\, dt = \kappa(\bar{\alpha}\lambda(1-x)) = Y(x).$$

With this interpretation of $Y(x)$ we can see that (257) and (258) summarize the change of events over a service period. $A_1(x)$ generates the number of NP messages present just before the end of a PSP. The number of P messages at the beginning depends on how many were left at the end of the preceding service period. If the preceding SP was a NPSP, the probability that it ended leaving $m \geq 1$ P messages is μb_{m1}. The number of NP arrivals in this case has generating function $Y^m(x)$ and the contribution to μa_{1n}, the probability that the PSP terminates with n NP messages waiting, is the coefficient of x^n in $\mu b_{m1} Y^m(x)$. Summing over $m \geq 1$ gives the contribution $\mu(B_1(Y(x)) - b_{01})$ to $\mu A_1(x)$. The alternative is that the preceding service period, whether NPSP or PSP, ended leaving an empty system. The next PSP must then start with the arrival of a P message (probability $\alpha\lambda p_0$) which generates NP messages according to $Y(x)$. Thus

$$\mu A_1(x) = \mu B_1(Y(x)) - \mu b_{10} + \alpha\lambda p_0 Y(x)$$

which is (257) when μ is divided out.

In conclusion, substitution of (257) and (258) into (255) and (256) gives

$$A(x,y) = \frac{y(B_1(y) - B_1(Y) + \alpha r p_0(y - Y))}{\alpha r(y - Y)(1/\alpha r Y - y)} \qquad (259)$$

$$B(x,y) = \frac{y(A_1(y) - A_1(Y) + \bar{\alpha} r p_0(y - Y))}{\bar{\alpha} r(y - Y)(1/\bar{\alpha} r Y - y)}. \qquad (260)$$

A partial check on (259) and (260) can be obtained by evaluating $A(1,1)$ and $B(1,1)$. These should sum to $1 - p_0 = r$. Writing

$$B_1(y) = \sum_{j=0}^\infty \frac{(y-Y)^j}{j!} B_1^{(j)}(Y)$$

we get from (259)

Sec. 5.3] A generating function approach to the state probabilities

$$A(x,y) = \frac{y\left(\alpha r p_0 + \sum_{j=1}^{\infty} \frac{(y-Y)^{j-1}}{j!} B_1^{(j)}(Y)\right)}{(1/Y - \alpha ry)} \qquad (261)$$

and so

$$A(1,y) = \frac{y\left(\alpha p_1 + \sum_{j=1}^{\infty} \frac{(y-1)^{j-1}}{j!} B_1^{(j)}(1)\right)}{(1 - \alpha ry)} \qquad (262)$$

since $Y(1) = 1$. $B^{(j)}(Y)$ denotes the jth derivative of $B_1(y)$ evaluated at $y = Y$. From (257) and (258) we also get

$$A_1^{(1)}(x) = (B_1^{(1)}(Y) + \alpha p_1) \, Y'(x)$$
$$B_1^{(1)}(x) = (A_1^{(1)}(\underline{Y}) + \bar{\alpha} p_1) \, \underline{Y}'(x) \, .$$

Now

$$Y'(x) = \bar{\alpha} r Y(x)/R$$

and therefore

$$Y'(1) = \bar{\alpha}r/(1 - \alpha r) , \qquad \underline{Y}'(1) = \alpha r/(1 - \bar{\alpha}r) \, .$$

Hence

$$A'(1) = \frac{\bar{\alpha}r}{1 - \alpha r}\left(\alpha p_1 + \frac{\alpha r(\bar{\alpha} p_1 + A'(1))}{1 - \alpha r}\right)$$

giving

$$A_1'(1) = \alpha \bar{\alpha} r^2 = B_1'(1) \qquad (264)$$

by symmetry.

It follows from (262) with $y = 1$ that

$$A(1,1) = \alpha r \, , \qquad B(1,1) = \bar{\alpha} r \, , \qquad (265)$$

symmetry giving the second.

Now

$$A(1,y) = \sum_{m \geq 1} A_m(1) y^m$$

and

$$A_m(1) = \sum_{n \geq 0} a_{mn}$$

so that $A(1,y)$ generates the probabilities $A_m(1)$ that the number X of P-messages waiting at *an arbitrary moment* during a PSP is m. Similarly $B(1,y)$ generates the probabilities $B_n(1)$ that the number Y of NP-messages waiting during a NPSP is n.

5.4 SYSTEM STATE PERFORMANCE

5.4.1 Formulae for mean and variance

It is of interest to develop expressions for the conditional means and variances. The means are given by

$$[dA(x,1)/dx]_{x=1} = A(1,1) E[Y|Z=1] = A(1,1) \frac{\bar{\alpha}r}{1-\alpha r} E[X|Z=1] \qquad (266)$$

$$[dB(1,y)/dy]_{y=1} = B(1,1) E[Y|Z=0] = \frac{\tfrac{1}{2}A''(1) + \bar{\alpha}r}{1-\bar{\alpha}r} \qquad (267)$$

$$[dA(1,y)/dy]_{y=1} = A(1,1) E[X|Z=1] = \frac{\tfrac{1}{2}B''(1) + \alpha r}{1-\alpha r} \qquad (268)$$

$$[dB(x,1)/dx]_{x=1} = B(1,1) E[X|Z=0] = B(1,1) \frac{\alpha r}{1-\bar{\alpha}r} E[Y|Z=0] . \qquad (269)$$

Here $A(1,1) = \alpha r$, $B(1,1) = \bar{\alpha} r$. The right-hand sides of (267) and (268) can be deduced immediately from (262) since the left-hand side, for example in (268), is the coefficient of y in $A(1,y)$. Equation (267) follows by symmetry. The relationship in (266) also comes directly from (261) by differentiating first with respect to x and then setting $x,y = 1$. Equation (269) is the symmetrical counterpart.

The variances can be obtained similarly, but are more cumbersome to evaluate explicitly.

The means require $A_1''(1)$ and $B_1''(1)$. Another differentiation of $A_1'(x)$ gives

$$A_1''(1) = \frac{2\alpha\bar{\alpha}^2 + \bar{\alpha}^2 r^2 B_1''(1)}{(1-\alpha r)^2} \qquad (270)$$

and by symmetry there follows

$$B_1'(1) = \frac{2\bar{\alpha}\alpha^2 r^3 + \alpha^2 r^2 A_1''(1)}{(1-\bar{\alpha}r)^2} .\qquad(271)$$

By elimination and symmetry we then have

$$A_1''(1) = \frac{2\alpha\bar{\alpha}^2 r^3(1-2\bar{\alpha}r+\bar{\alpha}r^2)}{(1-r)(1-r+2\alpha\bar{\alpha}r^2)}\qquad(272)$$

$$B_1''(1) = \frac{2\bar{\alpha}\alpha^2 r^3(1-2\alpha r+\alpha r^2)}{(1-r)(1-r+2\alpha\bar{\alpha}r^2)} .\qquad(273)$$

5.4.2 Numerical values and comments

Based on the above, Tables 26 and 27 have been constructed. Table 26 gives all four conditional expectations: $E[X|Z=1]$, $E[X|Z=0]$, $E[Y|Z=0]$, $E[Y|Z=1]$ for $r = 0.1, 0.5, 0.9, 0.95, 0.99$, and $\alpha = 0.1\,(0.2)\,0.9$. Naturally there are symmetries to be exploited, but it is convenient to be able to scan the whole range of values. Table 27 gives the unconditional mean values $E[X]$ and $E[Y]$ waiting and processing in the P and NP streams respectively. These last means were calculated from Table 26 by the weighted sums

$$\begin{aligned} E[X] &= A(1,1)\,E[X|Z=1] + B(1,1)\,E[X|Z=0] \\ E[Y] &= A(1,1)\,E[Y|Z=1] + B(1,1)\,E[Y|Z=0] . \end{aligned}\qquad(274)$$

This provides a check on the algebra. For a Peeping Tom would, at an arbitrary moment, observe on average what is shown by Table 27. The sum of the means is the mean number of messages in an $M/M/1$ system, whose value is known to be $r/(1-r)$. It will be seen that the pairs of columns in Table 27 do satisfy this relation, though it is not so easy to make the check algebraically without error.

Table 27 is more interesting as a basis for comment, and may be compared with Table 23 of Chapter 4 which gives similar means for head-of-the-line priority. Under heavy traffic the monstrously unbalanced waiting line of NP messages is avoided, and the load is fairly evenly spread with more or less equal misery for all. We shall turn to the system time later, but it is clear that shorter lines mean shorter waiting times. Thus, if a mistake in classifying a message as P or NP is made, when it should be the other, the damage is less serious. On the other hand genuine P messages which ought to arrive without delay are in fact delayed about half as much again as if no such regulatory mechanism were in operation. And here we should recall once more that this is not a true priority system but one which, like roundabouts rather than traffic lights at a road intersection, strives to facilitate the flow of vehicles by adapting to the needs of the traffic.

Table 26 — Values of $E[X|Z=1]$, $E[X|Z=0]$, $E[Y|Z=1]$, $E[Y|Z=0]$

$r = 0.1$

α	$E[X\|Z=1]$	$E[X\|Z=0]$	$E[Y\|Z=0]$	$E[Y\|Z=1]$	$E[X\|Z=1]$	$E[X\|Z=0]$	$E[Y\|Z=0]$	$E[Y\|Z=1]$
0.1	1.0112	0.0121	1.0999	1.0919	0.1330	1.1697	1.8670	0.5367
0.3	1.0334	0.0348	1.0777	0.0746	1.3347	0.3843	1.6653	0.5496
0.5	1.0556	0.0556	1.0556	0.0556	1.5	0.5	1.5	0.5
0.7	1.0777	0.0746	1.0334	0.0348	1.6653	0.5496	1.3347	0.3843
0.9	1.0999	0.0919	1.0112	0.0121	1.8670	0.5367	1.1330	0.1697

$r = 0.9$

α	$E[X\|Z=1]$	$E[X\|Z=0]$	$E[Y\|Z=0]$	$E[Y\|Z=1]$	$E[X\|Z=1]$	$E[X\|Z=0]$	$E[Y\|Z=0]$	$E[Y\|Z=1]$
0.1	4.0354	3.2990	6.9646	3.5919	8.7113	8.0512	12.2887	8.2309
0.3	5.0911	4.3119	5.9089	4.3937	10.0577	9.3096	10.9428	9.3539
0.5	5.5	4.5	5.5	4.5	10.5	9.5	10.5	9.5
0.7	5.9089	4.3937	5.0911	4.3119	10.9428	9.3539	10.0572	9.3096
0.9	6.9646	3.5919	4.0354	3.2990	12.2887	8.2300	8.7113	8.0512

$r = 0.99$

α	$E[X\|Z=1]$	$E[X\|Z=0]$	$E[Y\|Z=0]$	$E[Y\|Z=1]$
0.1	48.3754	47.7964	52.6243	47.8388
0.3	50.0304	49.3093	50.9696	49.3187
0.5	50.5	49.5	50.5	49.5
0.7	50.9696	49.3187	50.0304	49.3094
0.9	52.6243	47.8388	48.3754	47.7963

Table 27 — Unconditional mean values $E[X]$, $E[Y]$ of P and NP message numbers in system

α	$r = 0.1$		$r = 0.5$		$r = 0.9$		$r = 0.95$	
	$E[X]$	$E[Y]$	$E[X]$	$E[Y]$	$E[X]$	$E[Y]$	$E[X]$	$E[Y]$
0.1	0.0112	0.0999	0.1330	0.8670	3.0354	5.9646	7.7113	11.2887
0.3	0.0334	0.0777	0.3347	0.6653	4.0911	4.9089	9.0572	9.9428
0.5	0.0556	0.0556	0.5	0.5	4.5	4.5	9.5	9.5
0.7	0.0777	0.0334	0.6653	0.3347	4.9089	4.0911	9.9428	9.0572
0.9	0.0999	0.0112	0.8670	0.1330	5.9646	3.0354	11.2887	7.7113

α	$r = 0.99$	
	$E[X]$	$E[Y]$
0.1	47.3757	51.6243
0.3	49.0304	49.9696
0.5	49.5	49.5
0.7	49.9696	49.0304
0.9	51.6243	47.3757

5.5 SYSTEM TIME INCLUDING PROCESSING

5.5.1 Probability density functions and their Laplace transforms

Let W_p and W_{np} be the system times of P and NP messages respectively. Recalling that system time is the sum of waiting time and service (processing and transmission and decoding time), and writing $f(t) = \mu \exp(-\mu t)$, $f_n(t)$ its n-fold autoconvolution, $g_p(t)$, $g_{np}(t)$ the probability density functions of W_p and W_{np} respectively, we have

$$g_p(t) = p_0 f(t) + \sum_{n \geq 1} A_n(1) f_{n+1}(t) + \sum_{m \geq 0} b_{m1} k_{\bar{\alpha}}(t) * f_{m+1}(t)$$

$$\sum_{m \geq 0} b_{m2} k_{\bar{\alpha}}^{(2)}(t) * f_{m+1}(t) + \ldots + \sum_{m \geq 0} b_{mn} k_{\bar{\alpha}}^{(n)}(t) * f_{m+1}(t)$$

$$+ \ldots \quad (275)$$

Here $k_{\bar{\alpha}}^{(n)}(t)$ is the n-fold autoconvolution of an $\bar{\alpha}$-busy period density function. If $g_p(t)$ has Laplace transform $\gamma_p(z)$ then, with obvious notation,

$$\gamma_p(z) = \phi(z) [p_0 + A(1, \phi(z)) + B(\phi(z), \kappa_{\bar{\alpha}}(z))] \quad (276)$$

and by symmetry

$$\gamma_{np}(z) = \phi(z) [p_0 + B(1, \phi(z)) + A(\phi(z), \kappa_\alpha(z))] . \quad (277)$$

It suffices to explain (275). The p_0 term arises when an empty system is found upon arrival. If n P-messages are already in the system and a P-message is being processed (probability $A_n(1)$) the new message joins the P-queue, and the system time is the sum of the residual system time of the message being processed which has density $f(t)$ by the lack of memory property of the exponential distribution, the $n-1$ system times of the messages ahead in the queue and own system time. If NP messages are being processed and there are n present, the first component of the wait is the time to clear the NP queue. This is the sum of the residual busy period corresponding to the NP message being processed, which again by lack of memory has probability density function $k_\alpha(t)$, and the $n-1$ busy periods initiated by each of the remaining NP messages found, and taking place in sequence: if m (≥ 0) P-customers are present already, the new arrival has a further system time composed of the sum of $m+1$ processing times. The contribution is accordingly

$$b_{mn} f_{m+1}(t) * k_{\bar{\alpha}}^{(n)}(t) .$$

The properties of convolutions under the Laplace transformation and the definitions of $A(x,y)$ and $B(x,y)$ give (276). We can immediately check that $\gamma_p(0) = 1$, as should be the case.

5.5.2 Mean system time: numerical values and comments

$\gamma_p(z)$ can now be calculated directly in principle to give a measure of the reduction in value of α-stream information resulting from the alternating priority procedure. It can also be used to deduce the moments of W_p. Consider, for example, $E[W_p]$. If Little's formula holds we should have

$$E[W_p] = E[X]/\alpha\lambda ,$$

and we shall see later that this indeed is the case. But directly from (276) we obtain

$$E[W_p] = -\gamma_p'(0) =$$
$$= \frac{1}{\mu}\left[1 + \left\{\frac{dA(1,y)}{dy}\right\}_{y=1} + \left\{\frac{dB(x,1)}{dx}\right\}_{x=1} + \frac{1}{1-\bar{\alpha}r}\left\{\frac{dB(1,y)}{dy}\right\}_{y=1}\right]$$

giving

$$\mu E[W_p] = 1 + \alpha r\, E[X|Z=1] + \bar{\alpha} r\, E[X|Z=0] + \frac{\bar{\alpha}r}{1-\bar{\alpha}r} E[Y|Z=0] \qquad (278)$$

which can be expressed in various forms, one of which is

$$\mu E[W_p] = E[X] + \frac{(\tfrac{1}{2}A_1''(1) + \bar{\alpha}r)}{(1-\bar{\alpha}r)^2} + 1 , \qquad (279)$$

bearing a resemblance to Little's formula without actually being in the right form. It can be shown to reduce to Little's formula by showing that $E[X] - \alpha r E[W_p] = 0$, using (279) for $E[W_p]$ and setting $\mu = 1$. Equations (272) and (273) are needed.

Table 28 gives $E[W_p]$ in units of $1/\mu$ by direct calculation for $r = 0.9, 0.95, 0.99$, and we see readily that the values could have been obtained from Table 27 by use of Little's formula. In view of the symmetry there is no need to repeat this story for W_{np}, that is $E[W_p, \alpha] = E[W_{np}, \bar{\alpha}]$.

Table 28 shows that at these traffic levels there is a substantial variation in mean system time between $\alpha = 0.1$ and 0.9. In this sense the traffic with $\alpha > \bar{\alpha}$ appears to have a distinct advantage.

Table 28 — Mean P-class system time $E[W_p]$
(Time unit is mean time required to process a message)

α	$r = 0.9$	$r = 0.95$	$r = 0.99$
0.1	33.7266	81.1720	478.5428
0.2	20.8241	45.5726	245.3651
0.3	15.1522	31.7795	165.0855
0.4	11.9885	24.4826	124.4797
0.5	10	20	100
0.6	8.674	17.0116	83.6802
0.7	7.7919	14.9516	72.1062
0.8	7.2940	13.6068	63.6587
0.9	7.3637	13.2031	57.9397

To complete the comparison fairly it will be necessary to examine the variances too. They will suffer the disadvantage under heavy traffic of behaving like the inverse cube of a small quantity. Some values are given in Appendix C.

If the Laplace transform is to be used as a measure of degradation of information due to delay in arrival we need to be able to calculate $A(x,y)$ and $B(x,y)$ for arbitrary x and y. This topic is addressed in Appendix B, included for completeness. Table 29 gives values of $\gamma_p(z)$, $\gamma_{np}(z)$, and the weighted average $\alpha\gamma_p(z) + \bar{\alpha}\gamma_{np}(z)$.

Table 29 gives the value of message content degraded by a penalty factor equivalent to paying compound interest at 10% per time unit of delay in arrival. If the message were to arrive as soon as generated its value would be unity. Taking the P-stream on its own, we see from the column headed γ_p that P-traffic is degraded most for small α: that is to say, if P-traffic forms a small proportion of total traffic the value of its information is degraded more than if it formed the larger proportion of total traffic. A completely symmetrical statement can be made about NP-traffic and so we can generalize the statement to the point of obviousness:

Under an alternating priority system it is that class of message which forms the smaller proportion of total traffic whose value is degraded the more.

To evaluate the performance of the system as a whole we can consider the weighted mean $\alpha\gamma_p + \bar{\alpha}\gamma_{np}$, given in the third column of the table. In this case we see that the system performs worst when the two categories of traffic form equal proportions ($\alpha = \frac{1}{2}$), and best when there is a class of traffic which constitutes the larger proportion.

The advantage of the system compared with a no-priority single-server system can be gauged by comparing the $\gamma(z)$ that would apply in such a case. The values are given by $(1-r)/(z+1-r)$ at the foot of the table, and it is seen that the alternating priority system is in this sense more efficient.

To compare on the same basis with head-of-the-line priority see Table 25 of Chapter 4. At its best this degrades information to a lesser extent through late

Table 29 — Loss in value of information due to late arrival

$z = 0.1$

α	$r = 0.9$			$r = 0.95$			$r = 0.99$		
	γ_p	γ_{np}	$\alpha\gamma_p + \bar\alpha\gamma_{np}$	γ_p	γ_{np}	$\alpha\gamma_p + \bar\alpha\gamma_{np}$	γ_p	γ_{np}	$\alpha\gamma_p + \bar\alpha\gamma_{np}$
0.1	0.3661	0.5679	0.5477	0.2256	0.4249	0.4049	0.0626	0.1620	0.1521
0.3	0.4666	0.5331	0.5403	0.3185	0.4272	0.3946	0.1059	0.1646	0.1470
0.5	0.5331	0.5331	0.5331	0.3850	0.3850	0.3850	0.1405	0.1405	0.1405
0.7	0.5719	0.4666	0.5403	0.4272	0.3185	0.3946	0.1646	0.1059	0.1470
0.9	0.5679	0.3661	0.5477	0.4249	0.2256	0.4049	0.1620	0.0626	0.1521
	$\dfrac{1-r}{z+1-r} = 0.5$			$\dfrac{1-r}{z+1-r} = 0.33$			$\dfrac{1-r}{z+1-r} = 0.09$		

arrival, but if other factors to which attention has been drawn are taken into account the favourable balance may be reduced. Furthermore, the advantage of head-of-the-line priority wanes as traffic intensity r increases.

6

Modified Lanchester equations incorporating effects of information

6.1 INTRODUCTION

It is widely accepted that the possession of information about an enemy provides a military commander with strategic or tactical advantages. The history of warfare is filled with examples of the benefits that accrue from large advantages in information. Despite the empirical evidence, however, it is difficult to find convincing models of the effects of information in combat or, in particular, to demonstrate quantitatively the cumulative influence of small information advantages.

One of the simplest models of warfare is that provided by Lanchester's equations. The classical laws developed by Lanchester, [10], [17], [8], [18], apply to situations in which large aggregates of homogeneous forces oppose each other. In those circumstances, force strengths are assumed to be governed by first order differential equations that describe the rates of attrition inflicted by each side on the other. Replacement of forces by reserves can also be included easily in this format.

The classical Lanchester approach can be described briefly as follows.

Let the strengths of two opposing forces X and Y be described respectively by the time-varying quantities $x(t)$, $y(t)$. The time rates of change $\dot{x}$, $\dot{y}$ of these force strengths are then determined by the net balance between attrition and replacement. Thus

$$\dot{x} = (\text{Replacement})_x - (\text{Attrition})_x$$
$$\dot{y} = (\text{Replacement})_y - (\text{Attrition})_y \ .$$

A variety of results can be obtained from equations of this general form, depending on the assumptions made about the rates of attrition and replacement

Lanchester's original work did not consider replacement, and he used two different assumptions about attrition. The assumption of 'ancient warfare' holds that forces that lack the means to locate specific targets and wish to direct their fire against those targets must rely on random destruction of the enemy by area fire. In that case, the rate of attrition of, say, X depends on the product of force strengths xy, for the rate of fire by Y against X is proportional to y, and the rate of attrition per round is proportional to x, being in turn proportional to the density of targets within the field of fire.

Consequently, if both sides are engaged in ancient warfare, then the appropriate differential equations are

$$\dot{x} = -\gamma_1 xy$$
$$\dot{y} = -\gamma_2 xy \ , \tag{280}$$

where the constants γ_1, γ_2, represent the efficiencies of the fire of the two sides.

The alternative assumption, that of 'modern warfare', holds that the opponents can aim their fire with sufficient accuracy so that the rate of attrition does not depend on the density of targets. It depends only on the rate of fire, which is proportional to the strength of the firing force. If both sides are engaged in modern warfare, the appropriate differential equations are

$$\dot{x} = -\gamma_1' y$$
$$\dot{y} = -\gamma_2' x \ . \tag{281}$$

The solutions of (280) and (281) have quite different consequences. The transition from one set of solutions to the other, in essence the transition from ancient to modern warfare, is the central issue to be investigated in this chapter.

Conceptually, information about the enemy is one of the features that distinguish modern from ancient warfare. Under the assumption that the belligerents have weapons that can be directed accurately, the availability of accurate target-locating information is the factor that changes the mode of fighting from area fire to directed fire. A force with a large amount of target information can use directed fire; one with little information must rely on area fire.

In this chapter we formulate and solve a set of differential equations that describes this effect at a level of aggregation comparable to that of Lanchester's original equations. To do this, we make the following assumptions:

- Two forces, X and Y, oppose each other;
- Both provide replacements for lost units at specified constant rates;
- Y employs area fire;
- X employs directed fire when a large amount of information about Y is available; otherwise X employs area fire;
- The information itself is a dynamic variable that changes during the course of battle;
- The information is increased by the deliberate efforts of X;
- In the absence of deliberate and continuous efforts by X to obtain information, the

Introduction

amount of information available decreases to zero asymptotically because of target motion.

The addition of information as a dynamic battle variable is the major extension of the Lanchester formalism. It is recalled that the time-dependent, continuous variables x, y are respectively the force sizes of X and Y at time t after the beginning of battle.

Ideally, the information that enters the extended formalism should be the Shannon information about Y that is available to X. However, both the formulation and solution of Lanchester equations incorporating the correct form of the Shannon information appear unduly complex for a preliminary study. Consequently we introduce a proxy measure of information that shares some desirable properties of the Shannon information, but does not match it in all particulars.

Let z $(0 \leq z < \infty)$ represent the information variable. Then we make the following additional assumptions:

(1) $\dot{z}$ is an increasing function of x. Here we take $\dot{z}$ as simply proportional to x, although in reality a saturation effect may be present to limit the growth of z.
(2) $\dot{z}$ decreases linearly with z, so that in the absence of efforts by X, $z \sim e^{-vt}$. This is also a simplification of the way real tactical information decays, but it is a useful approximation.
(3) X does not necessarily switch abruptly from area fire to aimed fire. The transition may take place gradually as information is accumulated about Y's targets. Consequently the formalism should be able to treat an arbitrary transition in the type of fire.

These assumptions can be incorporated in a set of three differential equations, of which one describes the rate of change of information. Thus

$$\dot{x} = \lambda_1 - \gamma_1 xy \qquad (282a)$$

$$\dot{y} = \lambda_2 - \gamma_2 xy f - \gamma_3 x (1-f) \qquad (282b)$$

$$\dot{z} = \mu x - vz . \qquad (282c)$$

$f(z)$ is a non-increasing function of z, with the properties that

$$f(0) = 1$$

$$f(\infty) = 0 .$$

Equation (282a) states that the strength of the X force is being diminished at a rate $xy\gamma_1$ due to area fire from Y, and is being replaced at a constant rate, λ_1. Equation (282b) states that the strength of the Y force is being diminished at a rate $\gamma_2 fxy$ by area fire from X and at a rate $\gamma_3 x(1-f)$ by directed fire from X. The strength y of the Y force is being replaced at a fixed rate λ_2. Equation (282c) states that information is being obtained at a rate μx through the reconnaissance efforts of the X force. In the absence of any such efforts, the information decays in characteristic time v^{-1}.

This set of equations satisfies the minimum set of postulates that seem reasonable for exploring the effects of information. The subsequent sections of the chapter explore the solution of these equations analytically and numerically.

6.2 ANALYSIS

Before the full set of equations (282) is analysed, it is useful to consider the case of mixed ancient and modern warfare without reinforcement and without the effects of dynamically changing information. The reduced equations for that situation are

$$\begin{aligned} \dot{x} &= -\gamma_1 xy \\ \dot{y} &= -\gamma_2 xy - \gamma_3 x \end{aligned} \quad (283)$$

These can be integrated directly in the $x - y$ phase plane. We have

$$\frac{dy}{dx} = \frac{\gamma_3 + \gamma_2 y}{\gamma_1 y}, \quad (284)$$

which yields

$$x = x_0 + \frac{\gamma_1}{\gamma_2}(y - y_0) - \frac{\gamma_1 \gamma_3}{\gamma_2^2} \ln\left(\frac{\gamma_3 + \gamma_2 y}{\gamma_3 + \gamma_2 y_0}\right). \quad (285)$$

To determine the conditions for victory, we put $y = 0$. Then if

$$\begin{aligned} x|_{y=0} &> 0, \text{ X wins; conversely, if} \\ x|_{y=0} &< 0, \text{ Y wins.} \end{aligned}$$

The criterion for X to win is thus

$$\frac{x_0 \gamma_2}{y_0 \gamma_1} > 1 - \frac{\gamma_3}{\gamma_2 y_0} \ln\left(1 + \frac{\gamma_2 y_0}{\gamma_3}\right). \quad (286)$$

The limiting cases can be derived directly from this equation:

if $\quad \dfrac{\gamma_3}{\gamma_2} \to 0$, then $\quad \dfrac{x_0 \gamma_2}{y_0 \gamma_1} > 1 \quad (287)$

is the condition for the X-force to win, and if $\gamma_3/\gamma_2 \to \infty$, then

$$\frac{x_0\gamma_2}{y_0\gamma_1} > \frac{1}{2}\frac{\gamma_2}{\gamma_3}y_0 \tag{288}$$

is the corresponding condition.

Because the right-hand side of the inequality (286) is always less than 1, the effect of a non-zero γ_3 is always favourable to X. For fixed values of y_0, γ_1, and γ_2, X can win with a smaller initial force strength, x_0. With this point in mind, we return to the full three-dimensional set of equations.

Equations (282a,b,c) contain seven parameters that describe the rates and efficiencies that characterize a battle problem. The initial states of the variables, x_0, y_0, z_0, must also be included in the description of the problem. This presents the analyst with a ten-dimensional parameter space, within which the characteristics of the solutions, $x(t)$, $y(t)$, $z(t)$ must be explored. Since this space is prohibitively large for numerical analysis of the full range of possible cases, the first task is to attempt to reduce the number of parameters by rescaling the equations.

If we make the substitutions

$$\xi = x/a \qquad \eta = y/b \qquad \zeta = z/c \qquad \tau = t/d$$

then the scaling

$$a = v/\gamma_2 \qquad b = v/\gamma_1 \qquad d = 1/v$$

transforms equations (282a,b,c) into

$$\dot{\xi} = \pi_1 - \xi\eta \tag{289a}$$
$$\dot{\eta} = \pi_2 - \xi\eta f - \rho_3\xi(1-f) \tag{289b}$$
$$\dot{\zeta} = \varepsilon\xi - \zeta \tag{289c}$$

where the original seven parameters have been reduced to four:

$$\pi_1 = \lambda_1\gamma_2/v^2 \tag{290a}$$
$$\pi_2 = \lambda_2\gamma_1/v^2 \tag{290b}$$
$$\rho_3 = \gamma_1\gamma_3/\gamma_2 v \tag{290c}$$
$$\varepsilon = \mu/\gamma_2 c . \tag{290d}$$

$f(z)$ is now $f(c\zeta)$. c remains a free parameter to be chosen to simplify $f(c\zeta)$.

Equations (289a,b,c) will be studied in detail by numerical solution in the space of parameters π_1, π_2, ρ_3, ε, x_0, y_0, z_0. Before going to numerical solution, some further analysis is useful.

It is first important to enquire whether equations (289a,b,c) have steady-state solutions, and, if so, whether they are stable.

By putting $\dot{\xi} = \dot{\eta} = \dot{\zeta} = 0$, we get:

$$\xi^* = \zeta^*/\varepsilon \tag{291a}$$

$$\eta^* = \pi_1\varepsilon/\zeta^* \tag{291b}$$

$$f(c\zeta^*) = \frac{\pi_2\varepsilon/\rho_3 - \zeta^*}{\pi_1\varepsilon/\rho_3 - \zeta^*}. \tag{291c}$$

Equation (291c) is a functional equation for ζ^* which determines whether there is a steady solution. From the general form of $f(\zeta c)$, we know that f decreases monotonically from 1 to 0 as ζ goes from 0 to ∞. The right-hand side of (291c) has a zero at $\varepsilon\pi_2/\rho_3$ and a pole at $\varepsilon\pi_1/\rho_3$. If $\pi_1 > \pi_2$, the pole is to the right of zero, and if $\pi_2 > \pi_1$ the pole is to the left of zero. Thus, if we restrict ourselves to considering only positive values of ζ^*, there are at least two different situations, shown in Figs. 8, 9.

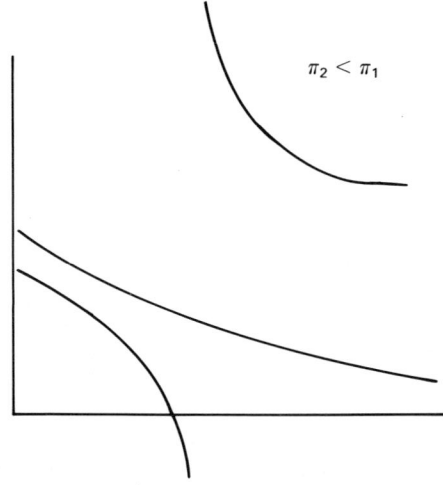

Fig. 8 — No solution.

It may also be possible that for some functions $f(c\zeta)$, there could be two solutions, shown in Fig. 10.

The latter obviously will depend on the specific form of the function $f(c\zeta)$.

Stability of the stationary points can be determined by linearizing equations (289a,b,c) and determining the time-dependence of the perturbations about the linear solutions. Specifically, if

$$\xi = \xi^* + \delta\xi \tag{292a}$$

$$\eta = \eta^* + \delta\eta \tag{292b}$$

$$\zeta = \zeta^* + \delta\zeta, \tag{292c}$$

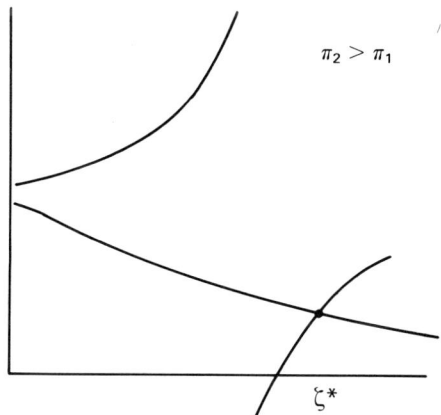

Fig. 9 — One solution at ζ^*.

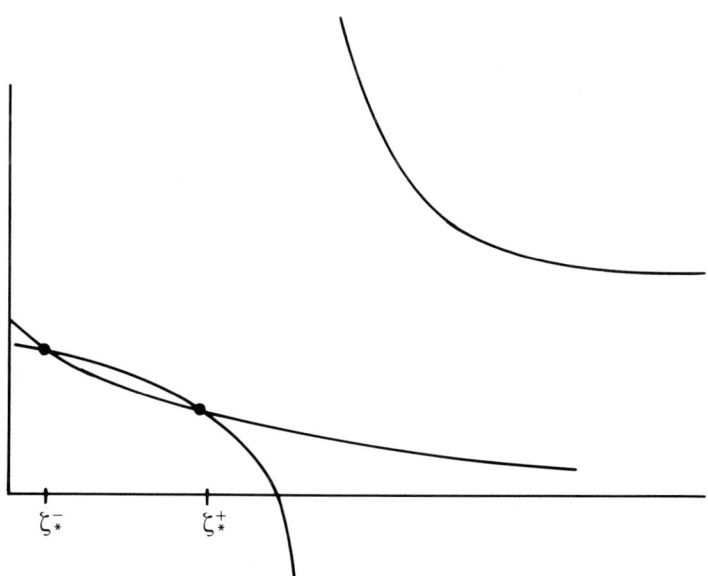

Fig. 10 — Two solutions.

then the perturbation equations are

$$\dot{\delta\xi} = -(\xi^*\delta\eta + \eta^*\delta\xi) \tag{293a}$$

$$\dot{\delta\eta} = -(\xi^*\eta^*\delta f + \xi^*f^*\delta\eta + \eta^*f^*\delta\xi) - $$
$$- \rho_3\delta\xi + \rho_3(f^*\delta\xi + \xi^*\delta f) \tag{293b}$$

$$\dot{\delta\zeta} = \varepsilon\delta\xi - \delta\zeta . \tag{293c}$$

If we assume that the perturbations are proportional to $e^{\Omega t}$, then Ω must satisfy a cubic equation derived from the determinant

$$\begin{vmatrix} \Omega + \eta^* & \xi^* & 0 \\ \eta^*f^* + \rho_3(1-f^*) & \Omega + \xi^*f^* & \xi^*\dfrac{\partial f^*}{\partial \zeta}(\eta^* - \rho_3) \\ -\varepsilon & 0 & \Omega + 1 \end{vmatrix} = 0 . \tag{294}$$

This is a cubic of the form

$$\Omega^3 + B\Omega^2 + C\Omega + D = 0 \tag{295}$$

where

$$B = 1 + \eta^* + \xi^*f^* \tag{296}$$

$$C = \eta^* + \xi^*f^* - \rho_3\xi^*(1-f^*) \tag{297}$$

$$D = -\varepsilon\xi^{*2}\frac{\partial f^*}{\partial \zeta}(\eta^* - \rho_3) - \xi^*\rho_3(1-f^*) . \tag{298}$$

The three roots of (295) have negative real parts, implying stability, if and only if

$$B > 0, \quad C > 0, \quad D > 0, \quad BC > D$$

simultaneously ([19], p. 109). Since B consists entirely of positive terms (for positive ζ^*), the first criterion is always satisfied. The latter three impose further constraints on the parameters of the problem if the system is to have stable stationary points.

With the use of (291), the inequalities that follow from equations (297) and (298) become

$$f^*(1 + 1/\rho_3) + \pi_1\varepsilon^2/\rho_3\zeta^{*2} > 1 \tag{299}$$

$$\zeta^*\frac{\partial f^*}{\partial \zeta}\left(1 - \frac{\pi_1\varepsilon}{\rho_3\zeta^*}\right) > 1 - f^* . \tag{300}$$

Analysis

The additional constraint that results from $BC > D$ is

$$\left(\frac{\pi_1\varepsilon^2}{\zeta^{*2}} + f^*\right)\left(f^*[1 + 1/\rho_3] + \frac{\varepsilon}{\rho_3\zeta^*} + \frac{\pi_1\varepsilon^2}{\rho_3\zeta^{*2}} - 1\right) >$$

$$> \varepsilon \frac{\partial f^*}{\partial \zeta}\left(1 - \frac{\pi_1\varepsilon}{\rho_3\zeta^*}\right). \tag{301}$$

Any further analysis of the inequalities (299) and (300) requires specification of the function $f(c\zeta)$.

Probably the simplest example that satisfies $f(0) = 1, f(\infty) = 0$ is

$$f(c\zeta) = \frac{1}{1 + c\zeta}. \tag{302}$$

With this choice of f, we can first determine the solution ζ^*. From (291),

$$\pi_2 - \frac{\rho_3\zeta^*}{\varepsilon} + \pi_2 c\zeta^* - \frac{c\rho_3\zeta^{*2}}{\varepsilon} = \pi_1 - \rho_3\frac{\zeta^*}{\varepsilon} \tag{303}$$

which has the solutions

$$\zeta^* = \frac{1}{2}\frac{\pi_2\varepsilon}{\rho_3}\left(1 \pm \sqrt{\left[1 - \frac{4\rho_3}{\varepsilon c\pi_2}\left(\frac{\pi_1}{\pi_2} - 1\right)\right]}\right). \tag{304}$$

The inequalities (299) and (300) become

$$\pi_1 c^2\varepsilon^2(1 + c\zeta^*) + c^2\zeta^{*2} > \rho_3 c^3\zeta^{*3} \tag{305}$$

$$\frac{c\varepsilon\pi_1}{\rho_3} > 2c\zeta^* + c^2\zeta^{*2}. \tag{306}$$

The inequality (301) is a bit more complicated. It can be expanded as

$$(\pi_1 c^2\varepsilon^2[1 + c\zeta^*] + c^2\zeta^{*2})(\pi_1\varepsilon^2 c^2[1 + c\zeta^*] + c^2\varepsilon\zeta^* +$$
$$+ c^2\zeta^{*2}(1 + \varepsilon c) - \rho_3 c^3\zeta^{*3}) > c^4\varepsilon\zeta^{*3}(\pi_1 c\varepsilon - \rho_3 c\zeta^*). \tag{307}$$

Before proceeding further, we must make some comments on (304). We want to find the number of real positive solutions for ζ^*. We find

$$\pi_1 < \pi_2 \Rightarrow 1 \text{ positive, 1 negative root}$$

$$\pi_2 < \pi_1 < \pi_2 + \frac{c\varepsilon\pi_2^2}{4\rho_3} \Rightarrow 2 \text{ positive roots}$$

$$\pi_2 + \frac{c\varepsilon\pi_2^2}{4\rho_3} < \pi_1 \Rightarrow 2 \text{ complex roots.}$$

We use (305) to (307) to check the stability of the real, positive roots.

Equation (306) yields simple unequivocal results. Using (291c) we can reduce (306) to

$$\frac{\pi_2\varepsilon}{\rho_3} > 2\zeta^* \tag{308}$$

or, with (304),

$$0 > \pm \sqrt{1 - \frac{4\rho_3}{c\varepsilon\pi_2}\left(\frac{\pi_1}{\pi_2} - 1\right)}. \tag{309}$$

consequently, $D > 0$ is always satisfied for ζ^{*-} and never satisfied for ζ^{*+}. Thus, the upper root is unstable. The lower root must be checked further against the inequalities (305) and (307) that follow from $C > 0$ and from $BC > D$.

Combination of (304) and (305) yields a general inequality that must be satisfied by the four parameters, π_1, π_2, ρ_3 and $c\varepsilon$. However, that expression is so complex and unwieldy that no useful criteria can be inferred. The same is true of (304) and (307). We can, however, make plausibility arguments for both stability and instability within the allowed range of ζ^{*-}.

In the parameter range

$$\pi_2 \leqslant \pi_1 \leqslant \pi_2 + \frac{c\varepsilon\pi_2^2}{4\rho_3} \tag{310}$$

ζ^{*-} is contained in

$$0 \leqslant \zeta^{*-} \leqslant \pi_2\varepsilon/2\rho_3. \tag{311}$$

If $\pi_1 = \pi_2 + \delta$, then

$$c\zeta^{*-} \sim \delta/\pi_2$$

and (305) is always satisfied as $\delta \to 0$. Inequality (307) is also satisfied as $\delta \to 0$. Thus $\zeta^{*-} \to 0$ is a stable root.

At the upper end of the range, with

Sec. 6.3] Numerical studies

$$\zeta^{*-} \cong \frac{\pi_2 \varepsilon}{2\rho_3} \; ; \quad \pi_1 \cong \pi_2 + \frac{c\varepsilon \pi_2^2}{4\rho_3} \; ,$$

the inequality (305) is approximately

$$\left(\frac{c\varepsilon\pi_2}{2\rho_3}\right)^2 (c\varepsilon - 1) + \left(\frac{c\varepsilon\pi_2}{2\rho_3}\right)\left(3c\varepsilon + \frac{1}{\rho_3}\right) + 2c\varepsilon > 0 \; . \tag{312}$$

For $c\varepsilon > 1$, this is always satisfied for positive values of the ratio $\frac{\varepsilon c \pi_2}{2\rho_3}$.

For $c\varepsilon < 1$, there is a large positive value of $\frac{\varepsilon c \pi_2}{2\rho_3}$ for which the inequality is violated and the root ζ^{*-} is unstable.

To summarize this section, the system of (289a,b,c) can demonstrate the following types of behaviour.

(a) For $\pi_1 < \pi_2$, there is one stationary point for $\xi, \eta, \zeta \geq 0$. That stationary point is unstable.

(b) For $\pi_2 < \pi_1 < \pi_2 + \frac{c\varepsilon \pi_2^2}{4\rho_3}$ there are two stationary points for $\xi, \eta, \zeta \geq 0$. The point with the larger value of ζ^* is always unstable. The point with the smaller value of ζ^* is stable for $\zeta^* \to 0$. If $c\varepsilon < 1$, at least one unstable region can be demonstrated if $\frac{\varepsilon c \pi_2}{2\rho_3}$ is sufficiently large.

(c) For

$$\pi_2 + \frac{c\varepsilon \pi_2^2}{4\rho_3} < \pi_1$$

there are no real stationary points.

The following section of this Chapter investigates the solutions numerically in the various regions.

6.3 NUMERICAL STUDIES

To study the system numerically, equations (289) were expressed as first order difference equations and integrated until either ξ or η assumed a negative value, or until a preset stopping time was reached. Various measures of combat effectiveness were then extracted from the numerical results:

(a) the winner, i.e. the side with residual forces when the opposing side had been reduced to zero;
(b) the duration of the battle;

(c) the losses by X (the side employing information to enhance its effectiveness);
(d) the exchange ratio

$$R = \frac{Y \text{ loss}}{X \text{ loss}}.$$

These measures of effectiveness were treated as dependent variables, and the influence of the parameters π_1, π_2, ρ_3, ε, ξ_0, η_0, and ζ_0 was investigated. Only a limited number of such relationships could be investigated. Consequently, emphasis was placed on the effect of varying the parameter ε, the efficiency of information-gathering by X. The analytical results of the previous section were used to determine the regions in which ε was to be varied.

Before the numerical results are examined, it is useful to develop some further insight as to the nature of the solutions by looking at the behaviour in the $\xi - \eta$ phase plane. Fig. 11 shows the $\xi - \eta$ plane for an arbitrary finite value of the information

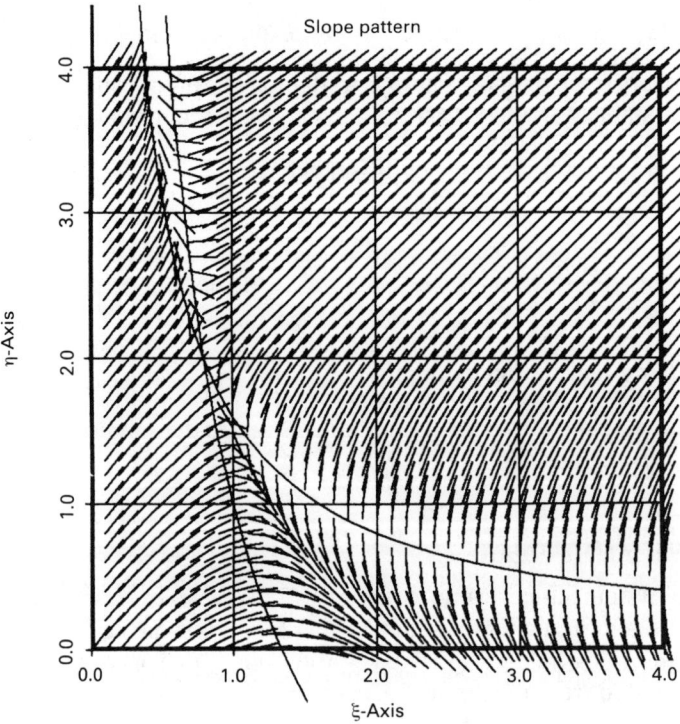

Fig. 11 — $\xi - \eta$ plane for an arbitrary finite value of the information parameter.

Sec. 6.3]   **Numerical studies**   147

parameter, ζ. The slope of the projection of the solution trajectory in the $\xi - \eta$ plane is given by

$$\frac{d\eta}{d\xi} = \frac{\pi_2 - \xi\eta f - \rho_3\xi(1-f)}{\pi_1 - \xi\eta} . \tag{313}$$

Consequently, the projected trajectories are horizontal along the curve

$$\eta = \frac{\pi_2}{\xi f} - \frac{\rho_3(1-f)}{f} , \tag{314}$$

and they are vertical along the curve

$$\eta = \pi_1/\xi . \tag{315}$$

Because ζ is a dynamic variable, the true solution trajectory does not remain in one $\xi - \eta$ plane, but moves in the perpendicular dimension as well, passing through a family of $\xi - \eta$ planes having the same general characteristics as Fig. 11.

The limiting cases for $f = 0$ and $f = 1$ are different. When no information is available, $\zeta = 0$ and $f = 1$, the two curves that divide the $\xi - \eta$ plane are

$$\eta = \pi_2/\xi \tag{316}$$

$$\eta = \pi_1/\xi \tag{317}$$

These are hyperbolae that converge as $\xi \to 0$ and $\xi \to \infty$ as shown in Fig. 12.

When perfect information is available $\zeta \to \infty$, $f \to 0$, and the corresponding curves are

$$\xi = \pi_2/\rho_3 \tag{318}$$

$$\eta = \pi_1/\xi \tag{319}$$

shown in Fig. 13.

An important qualitative feature of the solutions can be derived by inspection of Fig. 11. The slope $d\eta/d\xi$ is positive everywhere on the positive η axis. This means that for any starting point in the first quadrant (corresponding to physical reality) the solution can never approach the η axis for finite values of η. The only way $\xi \to 0$ is as $\eta \to \infty$. Consequently, Y can never win in finite time, only asymptotically. Thus one class of outcomes will be

$$\lim_{t \to \infty} x(t) = 0 .$$

$$\lim_{t \to \infty} y(t) \to \infty .$$

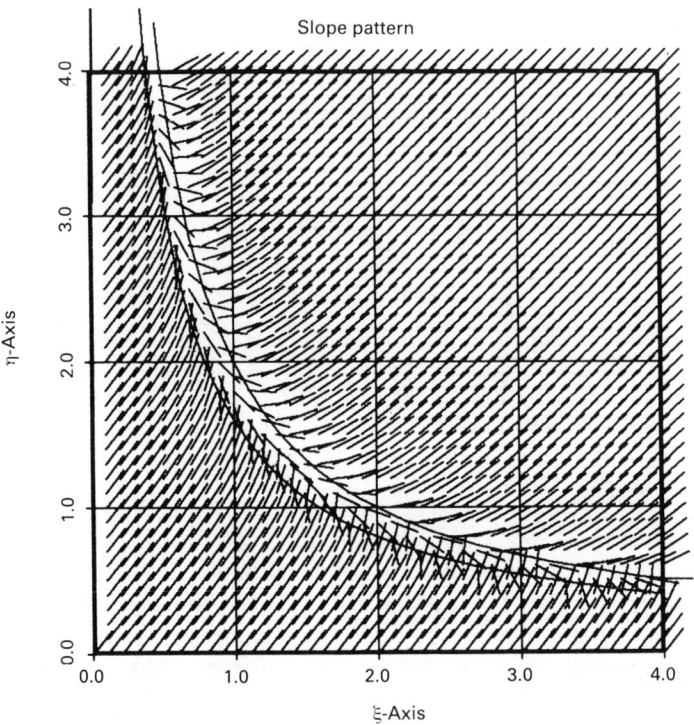

Fig 12 — $\xi - \eta$ plane with no information available.

The situation on the ξ axis is different. For $\xi < \dfrac{\pi_2}{\rho_3(1-f)}$, the slope is positive and the trajectories recede from the ξ axis. For $\xi > \dfrac{\pi_2}{\rho_3(1-f)}$ the slope is negative, and the trajectories approach the ξ axis. Consequently, there is a class of solutions for which $\eta \to 0$ for finite ξ. This means that X can win in finite time. The only exception to this is in the absence of information. Then $\zeta = 0$ and Fig. 12 governs the behaviour. In that case the only possible wins are asymptotic: either

$$x \to 0$$
$$y \to \infty ,$$

or

$$x \to \infty$$
$$y \to 0.$$

Sec. 6.3] Numerical studies

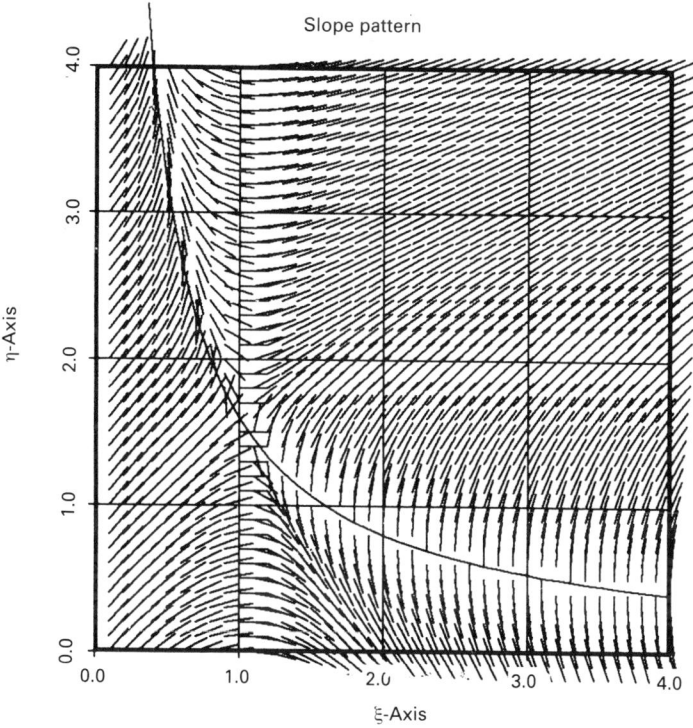

Fig. 13 — $\xi - \eta$ plane with perfect information.

Figs 14 to 19 will illustrate various modes of behaviour.

If the information variable were fixed so that X is employing a constant mix of aimed fire and area fire, then only four qualitatively different classes of solutions would be possible. Those can be inferred from inspection of Fig. 11. The four, which depend only on the initial values ξ_0, η_0, are:

(1) X wins: ξ increases monotonically; η first increases, then decreases to zero.
(2) X wins: ξ first decreases, then increases; η decreases monotonically to zero.
(3) Y wins: ξ decreases monotonically and approaches zero asymptotically; η first decreases, then increases asymptotically toward infinity.
(4) Y wins: ξ first increases, then decreases and approaches zero asymptotically; η increases monotonically and approaches infinity asymptotically.

When the information variable changes according to (289c), the classes of possible solutions are much richer. The four basic classes listed above still predominate; however, the effects of information modulate the basic solutions, change their shapes, add extra reversals to the fortune of battle, and in some cases change the outcome from one class to another.

Figs 14 and 15 show examples of classes (2) and (3) which are the most common because the starting points for solutions in those classes occupy most of the ξ, η plane. Each figure shows ξ, η, ζ as functions of t for 10^4 time steps.

The parameters are $\pi_2 = 2$, $\rho_3 = 3$, $\varepsilon = 0$, $\xi_0 = 2$, $\eta_0 = 3$, $\zeta_0 = 1$ for both figures. For Fig. 14, $\pi_1 = 1.8$; for Fig. 15, $\pi_1 = 2.2$. Thus, although X has some initial information, ζ_0, it decays to zero because $\varepsilon = 0$ and no new information is created. Consequently, the outcomes are ultimately determined by the replacement rates of the two forces. In Fig. 14, $\pi_2 > \pi_1$, and Y wins. In Fig. 15, $\pi_1 > \pi_2$ and X wins.

Fig. 16 shows a somewhat different situation. The parameters are the same as for Figs 14 and 15 except that $\pi_1 = 1.9$, only slightly below π_2. The strengths of both forces decrease rapidly initially. The initial information decays more slowly, and, for a brief time, the residual information is enough to turn the tide in favour of X. The strength of the X force increases slightly, but eventually the superior replacement rate of Y dominates, and Y wins asymptotically.

Figs 17 and 18 illustrate the effects of non-zero values of ε. The parameters for those cases are

$$\pi_1 = 1.6; \quad \pi_2 = 2; \quad \rho_3 = 3; \quad \xi_0 = 2; \quad \eta_0 = 3; \quad \zeta_0 = 1.$$

Thus Y has advantages in both initial force strength ($\eta_0 > \xi_0$) and replacement rate ($\pi_2 > \pi_1$). In Fig. 17, $\varepsilon = 0.6$. The strengths of X and Y both decrease initially, but then the effect of Y's greater replacement rate overwhelms any gains that X derives from better target information, so Y wins asymptotically. In Fig. 18, ε is increased to 0.8. That is enough to reverse the outcome of the battle. After an initial loss of force strength, X is able to regain the advantage and win in finite time. Note that the amount of information goes through two reversals in the course of battle.

In addition to those solutions for which there is a clearcut winner, there is also a stalemate solution as suggested by the stability analysis following the inequality (310). Fig. 19 shows such a solution for the parameters

$$\pi = 0.5, \quad \pi_2 = .4167, \quad \rho_3 = .4167$$
$$\varepsilon = 1.$$

These satisfy the inequality (310):

$$0.4167 < 0.5 < 0.5209.$$

The consequence is a stable stalemate with

$$\zeta^* \cong 0.2764$$
$$\xi^* \cong 0.2764$$
$$\eta^* \cong 1.809.$$

Sec. 6.3] Numerical studies

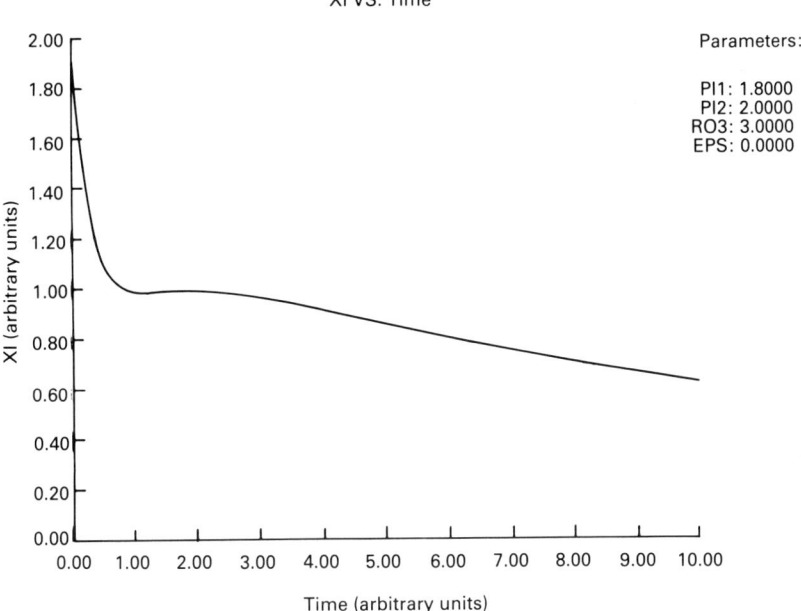

Fig 14a.

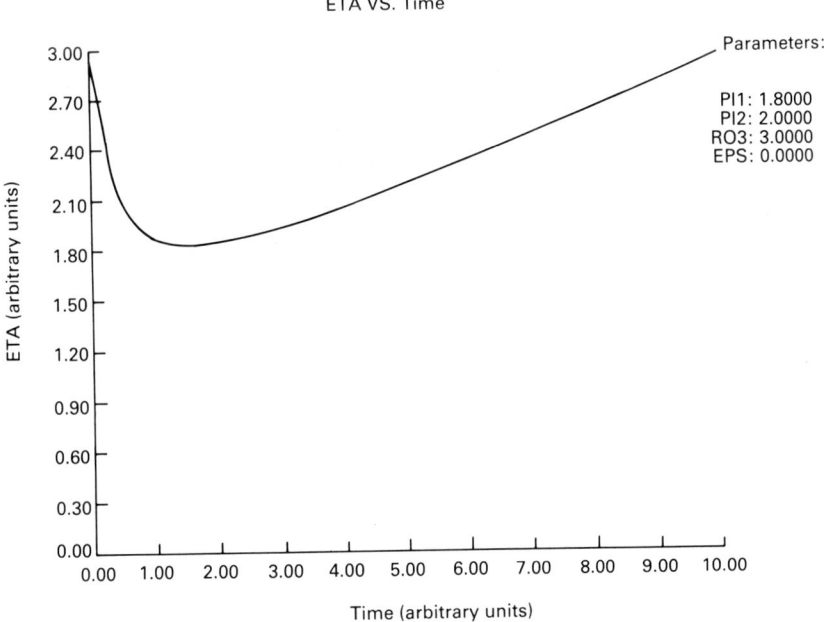

Fig 14b.

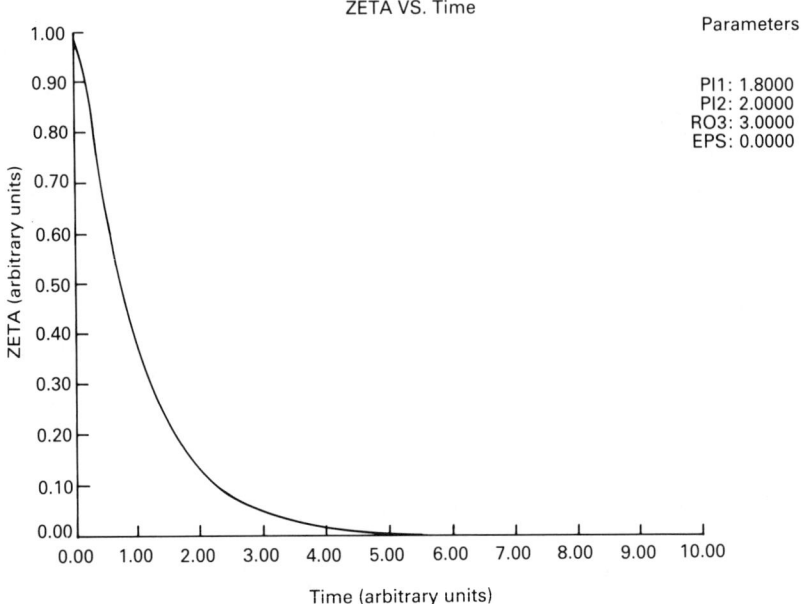

Fig 14c.

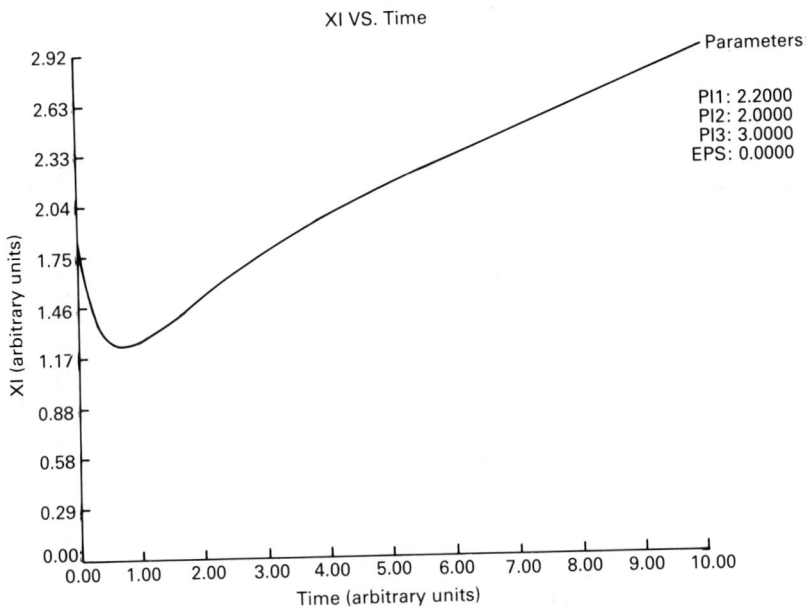

Fig 15a.

Sec. 6.3] **Numerical studies** 153

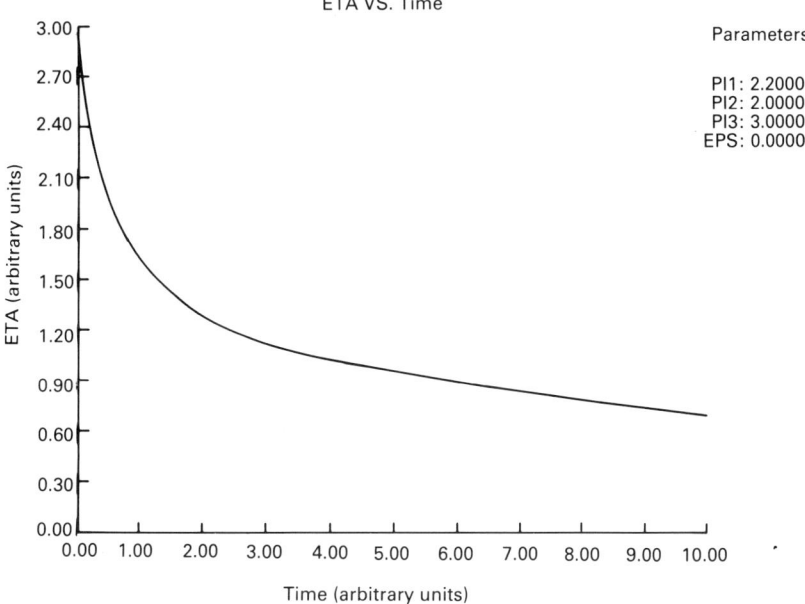

Fig 15b.

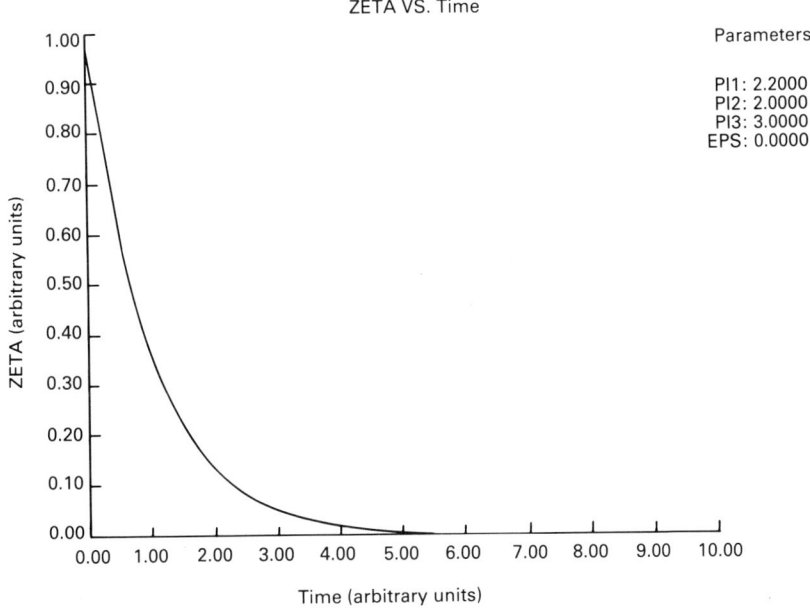

Fig 15c.

154 **Modified Lanchester equations** [Ch. 6

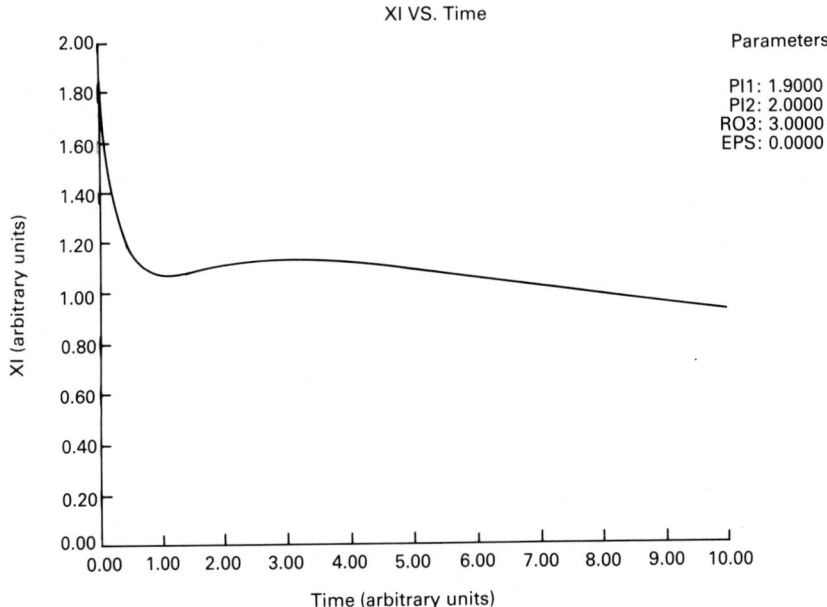

Fig 16a.

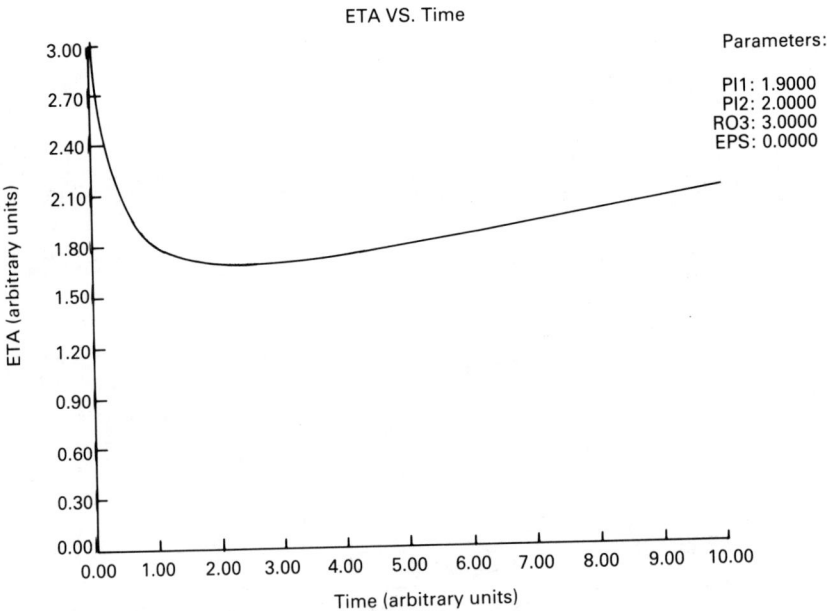

Fig 16b.

Sec. 6.3] Numerical studies

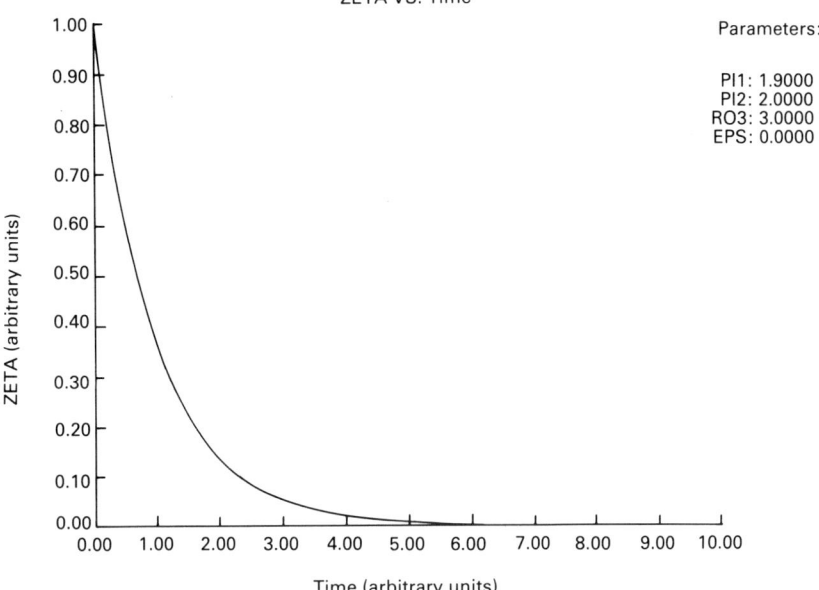

Fig 16c.

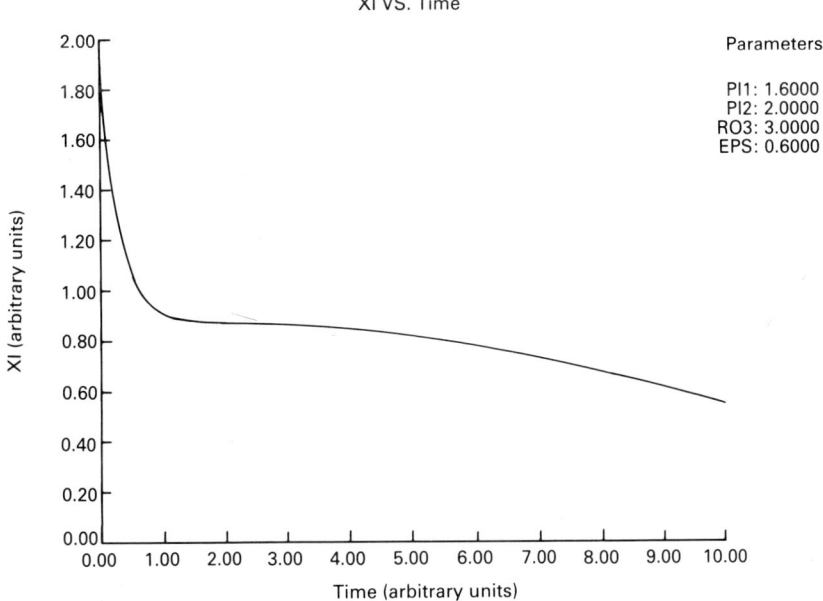

Fig 17a.

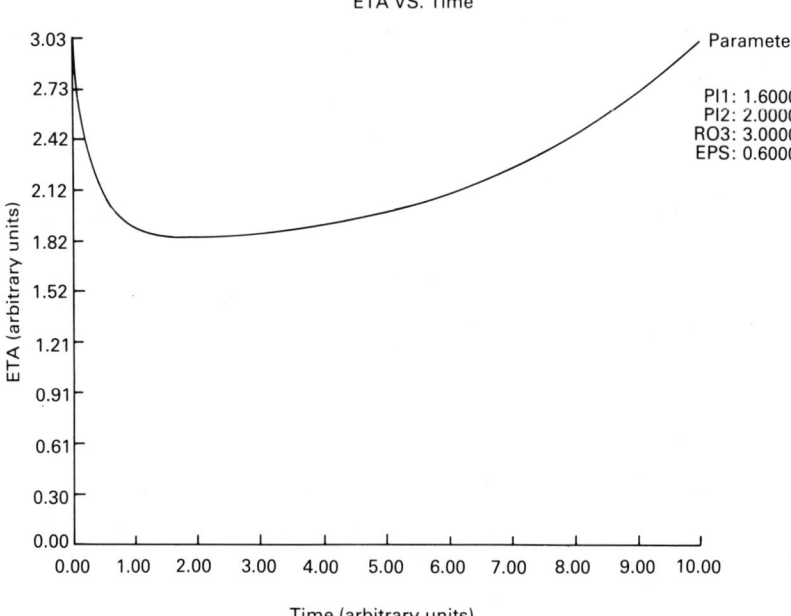

Fig 17b.

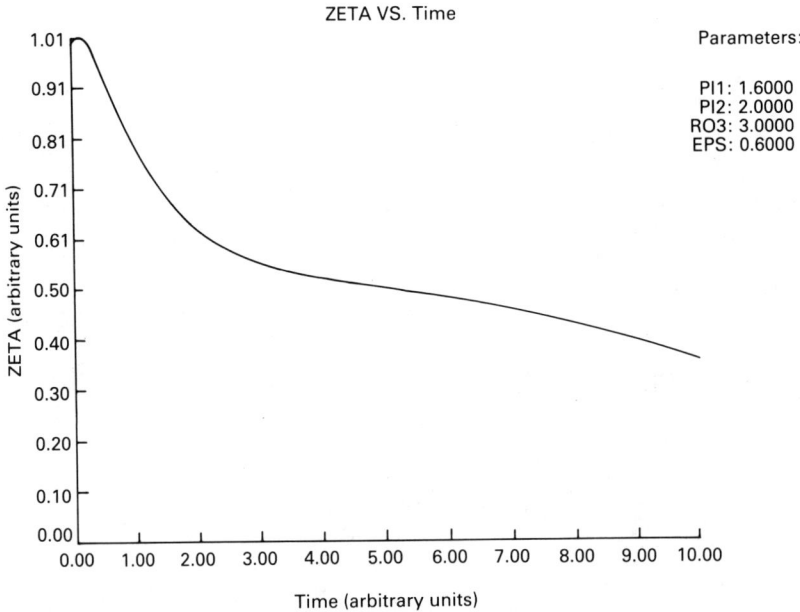

Fig 17c.

[Sec. 6.3] **Numerical studies**

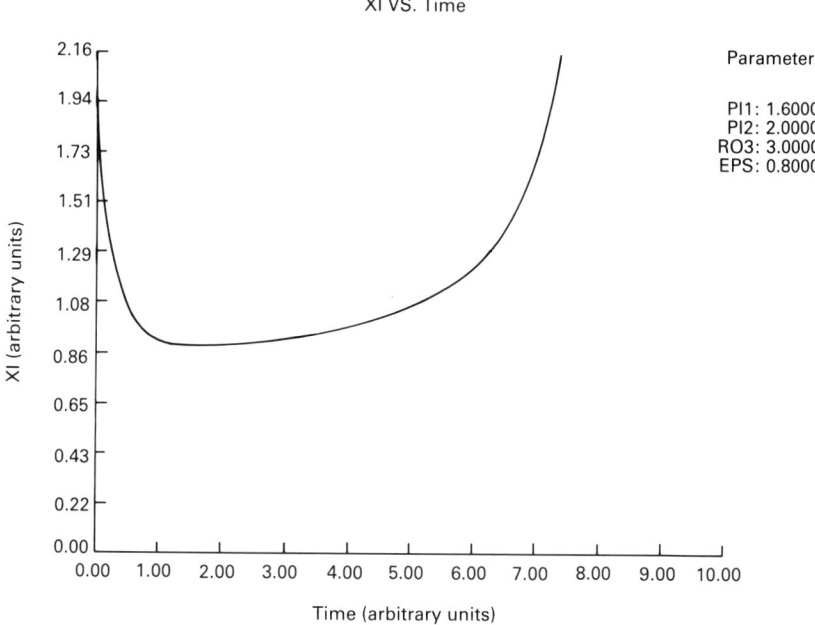

Fig 18a.

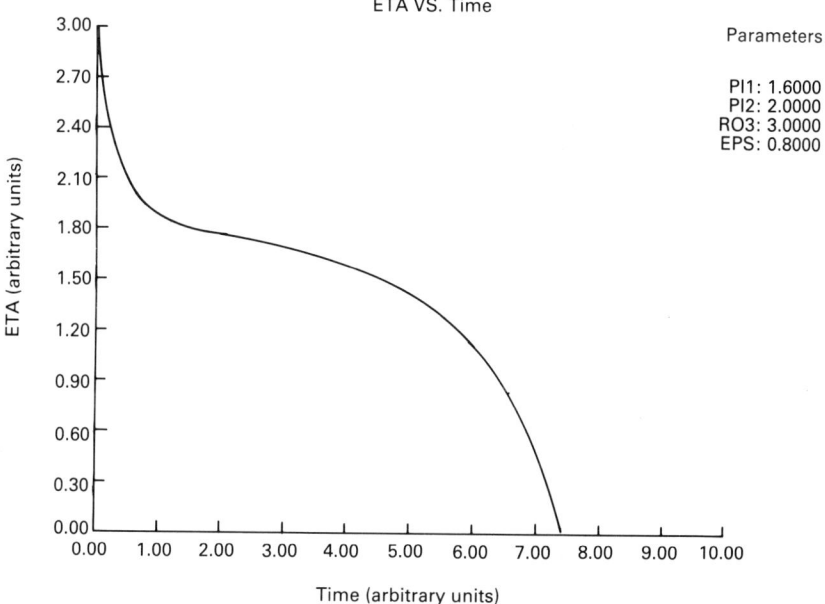

Fig 18b.

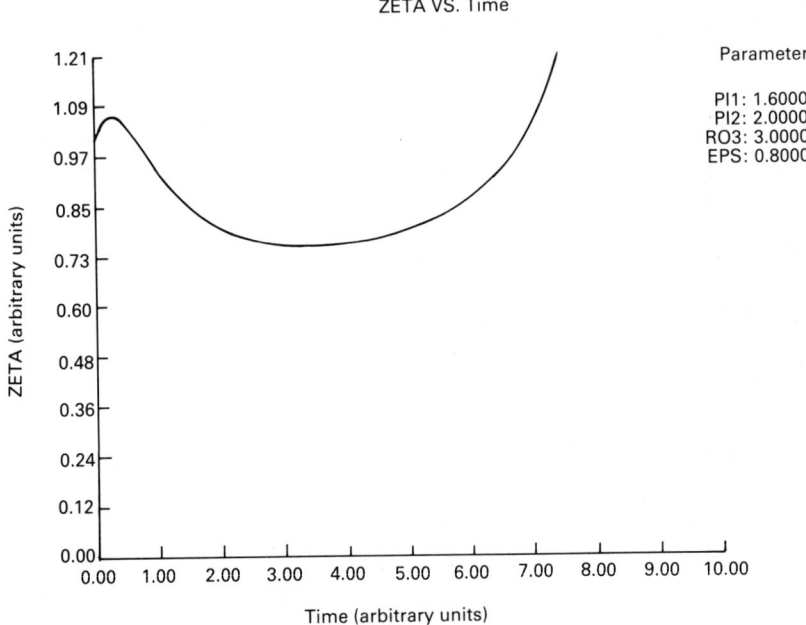

Fig 18c.

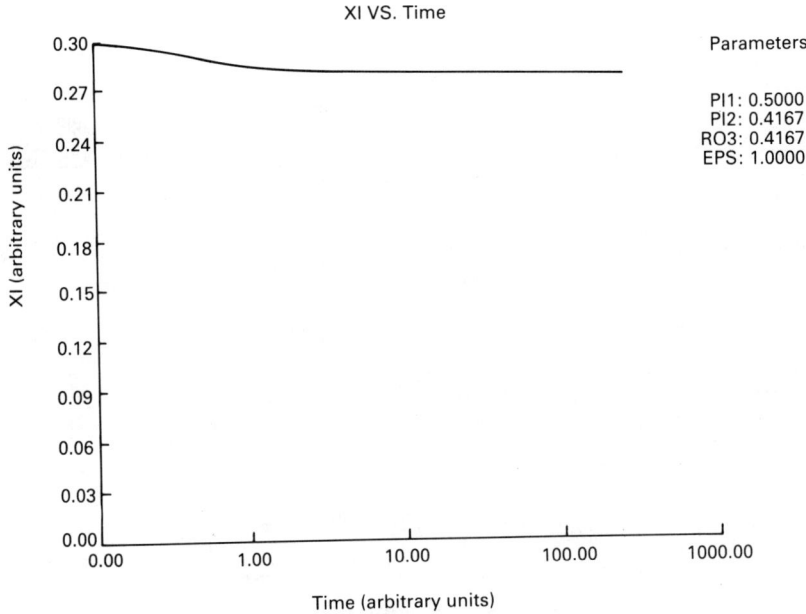

Fig 19a.

Sec. 6.3] Numerical studies

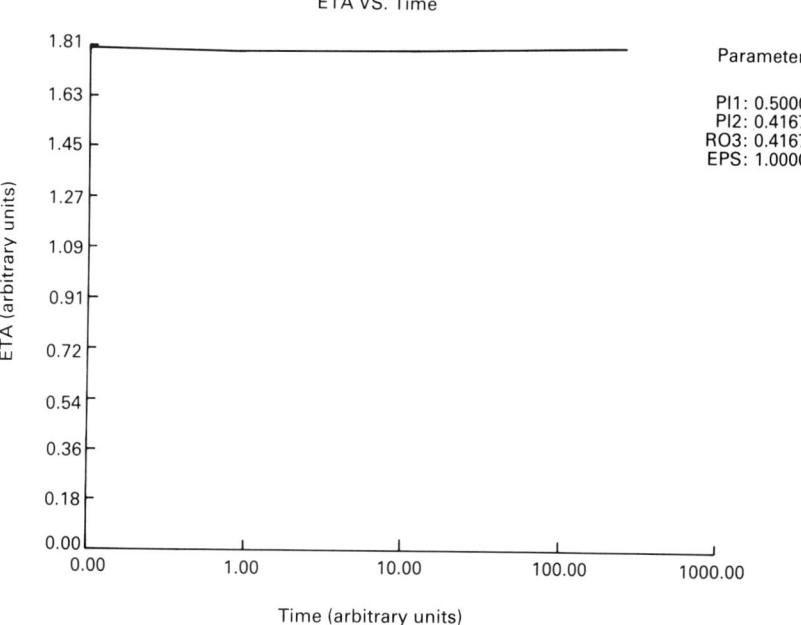

Fig 19b.

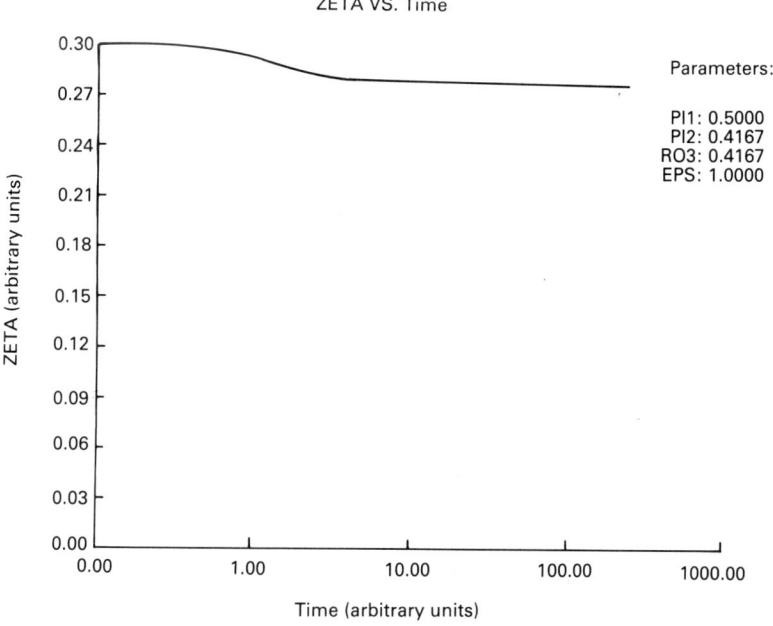

Fig 19c.

In this case, the availability of information enables X to hold Y to a stalemate despite a force ratio of about 6.5 in Y's favour. In the absence of this continuing production of information by X, Y would win asymptotically.

This is certainly suggestive of a situation in which a guerrilla force with excellent target intelligence can force a stalemate on a larger conventional army that must rely on area fire.

Diagrams like Figs 14 to 19 are useful for exploring the range of qualitative behaviours of the solutions. However, the influence of the parameters in equations (289) can best be understood by the quantitative comparison of a larger number of cases. Table 30 compares several measures of effectiveness as a function of ε for a

Table 30

$\pi_1 = 2.1 \quad \pi_2 = 2 \quad \rho_3 = 3$
$\xi_0 = 2 \quad \eta_0 = 3 \quad \zeta_0 = 1$

ε	Winner	Duration of battle	X_{loss}	R
0	X	$> 10^4$	20.91	1.05
0.4	X	2940	5.60	1.58
0.6	X	2548	4.97	1.62
0.8	X	2304	4.60	1.63
1.0	X	2208	4.40	1.67
3.0	X	1666	3.54	1.77
5.0	X	1536	3.30	1.82
8.0	X	1440	3.14	1.85

For battle durations greater than 10^4, X_{loss} and R are evaluated at 10^4.

case in which X has the advantage in rate of replacement. Table 31 makes a similar comparison for a case in which Y has the replacement advantage.

The exchange ratio, the losses by X, and the duration of battle are plotted in Figs 20 and 21.

Fig. 20 shows that although X always wins in that case, a small amount of added information makes a major difference to X: the battle is shortened, losses are drastically reduced, and the exchange ratio is improved. Fig. 21 emphasizes this and makes an additional point. In Fig. 21, for low values of ε, Y is the winner, X has large losses and an exchange ratio near parity. Small increases in ε do not change the situation. Y still wins, and the exchage ratio and losses by X are insensitive to added information. However, in the narrow region $0.6 < \varepsilon < 0.8$, major changes occur: X becomes the winner, X's losses are drastically reduced, and the exchange ratio becomes very favourable to X. Beyond this point, further increases in ε cause little change in the measures of effectiveness. Both the X losses and the exchange ratio are asymptotic to the values that would prevail if X were using aimed fire exclusively.

This demonstrates the existence of a threshold effect in the rate of production of

Sec. 6.3] **Numerical studies** 161

Table 31

$\pi_1 = 1.6 \quad \pi_2 = 2 \quad \rho_3 = 3$
$\xi_0 = 2 \quad \eta_0 = 3 \quad \zeta_0 = 1$

ε	Winner	Duration of battle	X_{loss}	R
0	Y	$>10^4$	17.66	1.03
0.2	Y	$>10^4$	17.64	1.04
0.4	Y	$>10^4$	17.60	1.08
0.6	Y	$>10^4$	17.44	1.15
0.8	X	7326	11.71	1.50
1.0	X	5390	8.72	1.56
3.0	X	2813	4.81	1.77
5.0	X	2425	4.22	1.86
8.0	X	2178	3.87	1.88

For battle durations greater than 10^4, X_{loss} and R are evaluated at 10^4.

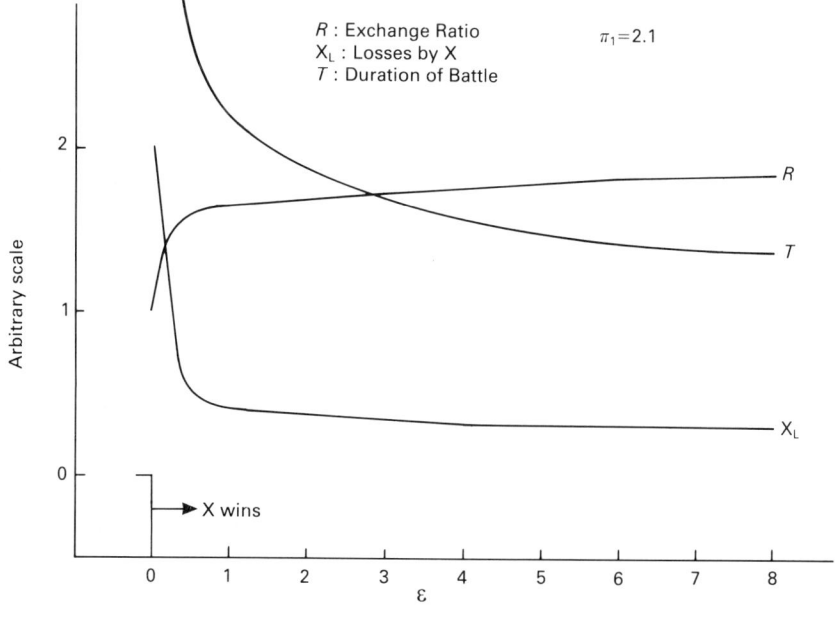

Fig. 20

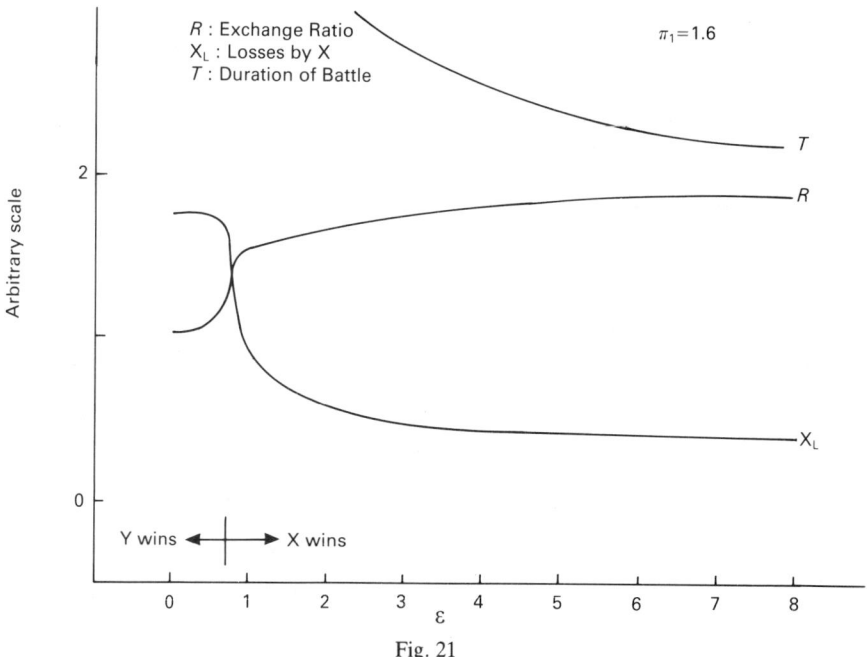

Fig. 21

information. Below the threshold, increases in ε produce very minor advantages for X. In the region of the threshold, small increases in ε reverse the trend of battle and provide X with major gains. Above the threshold, further large increases in ε produce ever-diminishing returns.

Equations (289) also include solutions for which information is counterproductive for X. This occurs if the scaled efficiency of aimed fire, ρ_3, is too low. Fig. 22 illustrates this effect. For that Figure X has the advantage in rate of replacement, $\pi_1 = 2.2$, $\pi_2 = 2$. However, $\rho_3 = 1.5$, half the value used in the previous examples. The Figure shows that X wins in the absence of information. However, even small rates of production of information cause X to switch to less efficient aimed fire, and as a consequence Y becomes the winner. The measure of effectiveness for X degrades slowly as ε increases. This is, of course, not a likely case. If the same weapons are used, the efficiency of aimed fire is greater than for area fire. However, if different weapons are used, and the efficiency of the aimed-fire weapons is sufficiently low, then the use of additional target information is indeed counterproductive.

6.4 SUMMARY

This chapter has explored the influence of dynamically changing levels of information on a battle described by a Lanchester-type model. To incorporate the effects of information, a third differential equation is added to two standard Lanchester equations. That third equation describes the rate of change of an information variable that can be replenished by the continuing efforts of the belligerent forces, but which decays to zero if it is not consciously replenished. The principal effect of

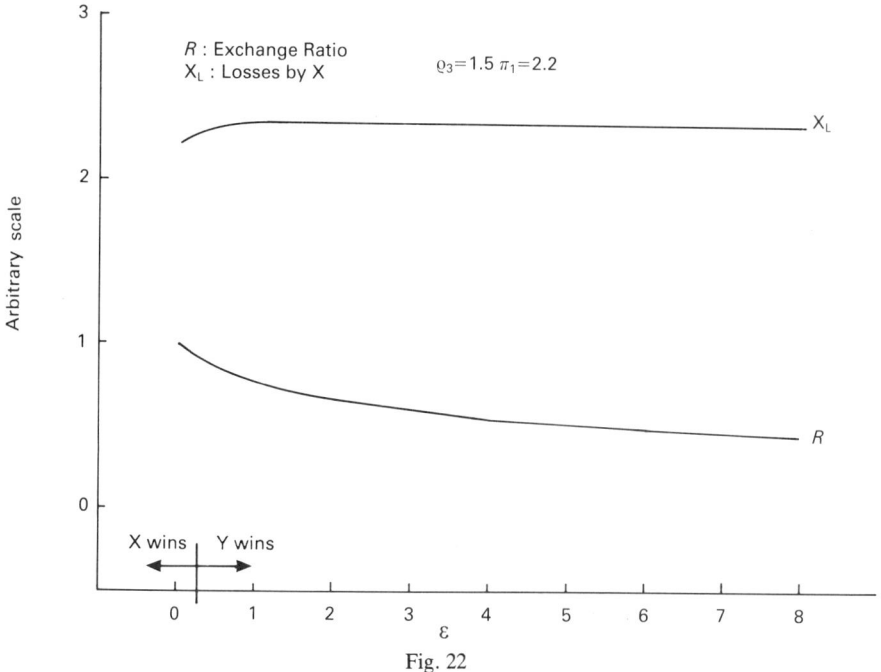

Fig. 22

the dynamic information level is to shift the mode of combat toward aimed fire by one force when the level of information is high, and toward area fire when the information level is low. The opposing force is assumed to use area fire at all times.

The addition of the information equation and its coupling with the standard Lanchester equations transform a relatively simple second order system into a complicated third order system. Among other features, the second order Lanchester system with replacements has at most one stationary solution that is always unstable, whereas the third order system can have two stationary points, one of which can be stable.

The third order system has been investigated theoretically and, within a small region of parameter space, numerically. Because the seven-dimensional parameter space is rather complex, the present investigation is only a beginning. Nevertheless, three important properties of information were identified as having significant consequences for combat in the Lanchester framework.

(1) New potential outcomes of battle are made possible. In the absence of information, there is always a winner and a loser. The force using only area fire can win asymptotically as $t \to \infty$, depending on initial conditions and rates of replacement; or, the force using a mix of area and aimed fire can win in finite time, also depending on initial conditions and rates of replacement.

When information is added, these two outcomes are still allowed, and are, in fact, observed in the majority of cases. However, there is an additional outcome, the stalemate, which is not allowed in the two-dimensional system. The stalemate occurs when a small force using good target information holds a much larger force in stable equilibrium. In that equilibrium, the rates of attrition and replacement are in balance for each force. It was not possible to investigate numerically the full range of conditions that may lead to stalemate. The theoretical analysis identifies the region in parameter space for which stalemate may occur, but does not identify the initial conditions that lead to stalemate.

(2) Under normal circumstances, information provides clearcut benefits to the force employing it. The duration of battle is decreased, losses are decreased, and the exchange ratio becomes more favourable as the rate of production of information is increased. The outcome of battle can also be reversed. Added production of information can change defeat into victory, and can compensate for unfavourable ratios of initial force strengths and rates of replacement.

It is a particularly important observation that the advantages of information manifest themselves rather abruptly. There is a threshold effect evident in a graph of, for example, the exchange ratio as a function of the rate of production of information. Far below the threshold, changes in information production have little influence on the exchange ratio. The same is true far above the threshold. However, in a narrow region around the threshold, small increases in the information production rate cause major improvement in the exchange ratio (and in other measures of effectiveness). It is thus very much in the interest of the information-employing force to understand where the threshold is and to ensure that the information production rate exceeds the threshold.

(3) Added information is not always beneficial. In circumstances for which the efficiency of aimed fire is low, additional information is counterproductive, and slavish adherence to a policy of aimed fire should be avoided.

The model used in this chapter is crude. Although it is complicated, any more realistic models of information in combat would be far more complicated. Nonetheless, they need to be studied.

Future studies should extend this work in several directions:

- to investigate the influence of initial conditions more thoroughly and to determine the basin of attraction to the stalemate solution;
- to consider the case in which Y uses all directed fire rather than all area fire;
- to use more realistic models of the information process, for example one that exhibits saturation behaviour;
- to treat the case in which Y also employs information, so that the situation is described by a fourth order system formally symmetric in x and y.

Appendices

APPENDIX A
TECHNICAL NOTES ON TEXT OF CHAPTERS 4 AND 5

The analysis in the text is based on the single-server queueing system usually designated in the literature by the notation $M/M/1$. The first M means that the arrival stream is Poisson and, therefore, that the random time intervals between successive arrivals are exponentially distributed: the mean arrival rate is usually denoted by λ. The second M means that the service times are exponentially distributed. The mean service time is usually denoted by μ^{-1}. The third symbol, in this case 1, designates the number of service channels. It is implicit here that the number of potential customers is infinite.

There is a wide choice of books describing this system and the associated stochastic processes of interest. This text uses results and notation based on [22].

The literature on priority queueing systems began sporadically in 1954 with the elegant paper by Cobham [21]. It has grown considerably, as is the fashion in queueing theory. The book by Jaiswal [14] is probably the first connected account, but although it too is elegant it is couched in very general terms and is somewhat academic in its orientation. Nevertheless some of the results we have derived independently are to be found in [14] as special cases. Indeed, we do not claim originality for anything in the text except that it has been developed specifically and independently in the context of a potential communication application. The type of priority described in Chapter 4 is called 'head-of-the-line' in the literature. There are many other types, one of which is 'pre-emptive', meaning that when a priority customer arrives he can interrupt the service of any non-priority customer who may be receiving attention. This did not seem appropriate in the context of communication. In the text we have considered only two classes of priority, but obviously a multiclass system is possible. Interesting in the message context is a system called 'alternating priority'. In this case, when one priority class is exhausted the service turns to the other class and serves it to exhaustion. This would tend to smooth out system times, but one wonders how much better it might be than the first come, first served discipline often adopted for non-priority systems. Some analysis is given in Chapter 5.

Conway, Maxwell, and Miller [23] is a masterful and interesting text aimed at sequencing and processing jobs in an industrial context. Kleinrock [24] is a text aimed primarily at computer systems and networks, and is full of wisdom and common sense. It is of potential interest in the study of communication systems.

Reference is made in Section 4.3 to the 'method of balance'. This is a convenient and easily comprehended method for the derivation of the state probabilities for simple systems with exponential characteristics in equilibrium. In the normal M/M/1 system, for example, the system state is a random variable N capable of taking non-negative integer values n with probability p_n. One might think of the total probability, whose mass is unity, as being divided into boxes labelled $n = 0, 1, 2. . .$ If the dynamics of arrival and service cause probability to flow, as it were, out of a box, this flow must, if the system is in equilibrium, be balanced by a total flow in of the same amount. The flow into and from other than adjacent boxes has negligible probability. As an example consider box n ($n \geq 1$) and its two neighbours labelled $n - 1$ and $n + 1$. The mass p_n of probability in box n will increase if there is flow from boxes $n - 1$ and $n + 1$. Flow from box $n - 1$ occurs at rate λ and its amount per unit time is λp_{n-1} due to arrivals, and from box $n + 1$ comes μp_{n+1} per unit time due to departures. Flow out of box n is likewise of amount λp_n into box $n + 1$ due to arrivals, and μp_n into box $n - 1$ due to service completions. Thus, for equilibrium, we must have a balance equation

$$(\lambda + \mu)p_n = \lambda p_{n-1} + \mu p_{n+1}, \quad (n \geq 1) .$$

For $n = 0$, a moment's thought shows that the balance equation is

$$\lambda p_0 = \mu p_1 .$$

The whole system can be seen easily to have solution $p_n = (1 - r)r^n$ ($r = \lambda/\mu$). Try substitution if all else fails. The condition for equilibrium is $r < 1$. Otherwise all $p_n = 0$ and with probability one the system contains infinitely many clients. The result above, quoted in Section 4.3 of the text, at (164) and (172), is the equation we have indicated above.

Reference is made in Section 4.7 to the very useful and versatile Little's formula. This links mean system time $\hat{W}$ with mean system state $\hat{N}$. If we were to imagine customers entering a large room upon arrival and moving through it one after another in a line until their service is completed we could see that the average time a typical customer spends in the room is $\hat{W}$, and, moreover, that at any one time the mean number of customers in the room is $\hat{N}$. The mean spacing between customers is $\hat{T}$, the mean time between successive arrivals, and thus we would expect to find that

$$\hat{N} \hat{T} = \hat{W} .$$

This is Little's formula, result, or theorem which can be given a rigorous proof under less restrictive conditions.

We have checked that (194) and (196) are obtainable as special cases of results in [14]. Jaiswal does not appear to derive formulae for the variances. However, it is extremely unlikely that someone has not carried out an analysis similar to ours. We reserve to ourselves the particular slant given by the comments and observations to which the analysis leads.

The busy period is prominent in the analysis of service systems with priorities. In the context of a single-server system a busy period T is defined to begin when the service is relieved from idleness by an arrival, and continues until it becomes idle again. Let $k(t)$ be the probability density function of T, and suppose that service time S has probability density function $f(t)$. We continue to assume in what follows that arrivals form a Poisson stream with mean rate λ. If no arrivals take place during the service of the first customer, $T = S$: otherwise, T is the sum of S and the continuous periods of activity, one following the other, generated by the successively arriving customers in the first service time. Thus

$$k(t) = e^{-\lambda t}[f(t) + \sum_{n \geq 1} \frac{(\lambda t)^n}{n!} f(t) * k^{(n)}(t)]$$

where, as usual, $*$ denotes convolution of the functions it separates and $^{(n)}$ denotes autoconvolution n times. The Laplace transform $\kappa(z)$ of $k(t)$ can be expressed in terms of the Laplace transform $\phi(z)$ of $f(t)$ by

$$\kappa(z) = \phi(z+\lambda) + \sum_{n \geq 1} \frac{[-\lambda\kappa(z+\lambda)]^n}{n!} \frac{d^n}{dz^n} \phi(z+\lambda) .$$

The right-hand side is recognized to be the expansion of $\phi(z + \lambda - \lambda\kappa(z+\lambda))$ so that $\kappa(z)$ is the smallest positive root of the functional equation

$$\theta = \phi(z + \lambda - \lambda\theta) ,$$

and functions of this root can, if necessary, be expanded by Lagrange's Theorem.

When $f(t) = \mu e^{-\mu t}$, as has been the case in our analysis, $\phi(z) = \mu/(\mu + z)$ and so $\kappa(z)$ is the smaller root of

$$\theta = \mu/(z + \lambda + \mu - \lambda\theta) ,$$

that is, of the quadratic equation

$$\lambda\theta^2 - (z + \lambda + \mu)\theta + \mu = 0 .$$

Let $Y(x)$ be the generating function of the number of arrivals of $\bar{\alpha}$-customers during an α-busy period. Then

$$Y(x) = \int_0^\infty \exp(-\bar{\alpha}\lambda t) \, \kappa_\alpha(t) \sum_{n \geq 0} \frac{(\bar{\alpha}\lambda xt)^n}{n!} \, dt = \kappa_\alpha[\bar{\alpha}\lambda(1-x)] \, .$$

Thus $Y(x)$ is the smallest root of the equation

$$0 = \phi[\bar{\alpha}\lambda(1-x) + \alpha\lambda - \alpha\lambda\theta] = \phi(\lambda - \bar{\alpha}\lambda x - \alpha\lambda\theta) \, .$$

Suitable functions of θ can be expanded by using Lagrange's Theorem.

The comments on management are a collection of observations well known to queueing theorists, made here to underline the importance of certain elementary management tasks which are fundamental. We claim to ourselves the use of the Laplace transform as a measure of the degradation in the value of the processed product on arrival at its destination. When it is valid this is an unlooked-for spin-off from a device used mainly as an analytical tool.

APPENDIX B
CALCULATION OF $A(x,y)$, $B(x,y)$ FOR $x,y \, \varepsilon (0,1)$

Reference to (255), (256) or to (259), (260) shows that $A(x,y)$ and $B(x,y)$ are completely determined if we know $A_1(x)$, $B_1(x)$. These are given by (257) and (258). Substituting from (258) into (257) and vice versa we obtain the pair

$$A_1(x) = A_1(\underline{Y}(Y(x))) - p_1(1 - \bar{\alpha}\underline{Y}(Y(x)) - \alpha Y(x))$$

$$B_1(x) = B_1(Y(\underline{Y}(x)))p_1(1 - \alpha Y(\underline{Y}(x)) - \bar{\alpha}\underline{Y}(x)) \tag{B.1}$$

for which we have used $a_{10} + b_{01} = p_1$.

Define sequences u_n, v_n as follows:

$$u_0 = x, \, v_1 = Y(u_0), \, u_2 = \underline{Y}(v_1), \ldots u_{2n} = \underline{Y}(v_{2n-1}), \, v_{2n+1} = Y(u_{2n}) \, . \tag{B.2}$$

It is easily seen that $\{u_{2n}\}$ and $\{v_{2n+1}\}$ are monotonically increasing sequences in n for $0 \leq x < 1$ provided that $r < 1$ (condition also for statistical equilibrium). For if $x = Y(x)$ has a root it satisfies $\alpha r x^2 - (1 + r - \bar{\alpha}rx)x + 1 = 0$, that is $rx^2 - (1+r)x + 1 = 0$ of which the roots are 1 and $1/r$. Now $Y(0)$, $\underline{Y}(0)$ are both positive and less than unity, and so $Y(x) - x$ and $\underline{Y}(x) - x$ are positive when $x = 0$. Both $Y(x)$ and $\underline{Y}(x)$ increase steadily as x increases, and accordingly $Y(x) - x$, $\underline{Y} - x$ remain positive over $0 \leq x < 1$, decreasing to zero for the first time when $x = 1$ provided that $r \leq 1$. Thus the sequences $\{u_{2n}\}$ and $\{v_{2n+1}\}$ both tend to unity for every x in $0 \leq x \leq 1$.

The first equation of the pair (B.1) then gives, successively,

$$A_1(u_0) = A_1(u_2) - w_0$$
$$A_1(u_2) = A_1(u_4) - w_2$$
$$A_1(u_4) = A_1(u_6) - w_4$$
$$\ldots\ldots\ldots\ldots\ldots\ldots$$
$$A_1(u_{2n}) = A_1(u_{2n+2}) - w_{2n}$$

where

$$w_{2n} = p_1(1 - \bar{\alpha} u_{2n+2} - \alpha v_{2n+1}) \;.$$

Addition gives

$$A_1(u_0) = A_1(u_{2n+2}) - \sum_{i=0}^{n} w_{2i} \;.$$

The series has to be convergent since $u_{2n} \to 1$, and $A_1(1)$ is finite by its very nature. The practical implication is that if the series is computed using enough terms to give a prescribed accuracy we shall have $A_1(u_0)$ to the same accuracy since u_{2n+2} will be unity to the same accuracy. All that accordingly remains is to calculate $A_1(1)$. This can be done in the same way by starting with $u_0 = 0$. Then $A_1(0) = a_{10} = \alpha p_1$ since by balance we must have $\alpha \lambda p_0 = \mu a_{10}$ and $p_1 = r p_0$. Accordingly

$$A_1(1) = \alpha p_1 + \lim_{n \to \infty} \sum_{i=0}^{2n} w_{2i} \;. \tag{B.3}$$

B(1) can be obtained similarly, or by using Eq. 246, namely

$$B_1(1) = A_1(1) + (\bar{\alpha} - \alpha) p_1 \;. \tag{B.4}$$

This is the basis for the calculations made for Table 29.

APPENDIX C
SOME VALUES OF THE STANDARD DEVIATIONS FOR ALTERNATING PRIORITY

A considerable amount of labour is needed to evaluate the second order statistics, and the details will not be given here. Some numerical values are, however, given in Table C.1 to facilitate comparison with the head-of-the-line priority system analyzed

Table C1 — Standard deviations for alternating priority

$r = 0.1$				0.5			0.9		
α	σ_X	σ_Y	$\sigma_{W_{np}}$	σ_X	σ_Y	$\sigma_{W_{np}}$	σ_X	σ_Y	$\sigma_{W_{np}}$
0.1	0.1064	0.3315	1.1093	0.4104	1.2606	1.8887	5.0065	6.2110	7.2624
.3	.1859	.2893	1.1092	.7082	1.0484	1.8838	5.8800	5.9347	9.6980
.5	.2422	.2422	1.1140	.8880	.8880	2.1512	5.9706	5.9706	13.788
.7	.2893	.1859	1.1239	1.0480	.7082	2.7247	5.9347	5.8800	21.662
0.9	0.3315	0.1064	1.1392	1.2606	0.4104	3.7653	6.2110	5.0065	52.375

$r = 0.5$				0.99					
α	σ_X	σ_Y	$\sigma_{W_{np}}$	σ_X	σ_Y	$\sigma_{W_{np}}$			
0.1	11.577	12.079	14.391	63.546	62.835	76.676			
.3	12.372	12.286	20.737	64.048	63.835	110.98			
.5	12.410	12.410	29.705	64.037	64.037	159.94			
.7	12.286	12.372	46.913	63.835	64.048	256.58			
0.9	12.079	11.577	120.19	62.835	63.546	680.62			

in Chapter 4.

The table gives standard deviations σ_X and σ_Y of the numbers X and Y of P- and NP-messages in the system, and the standard deviation $\sigma_{W_{np}}$ of W_{np}, the NP-message system time. σ_{W_p} can be inferred by symmetry since $\sigma_{W_p}(\alpha) = \sigma_{W_{np}}(\bar{\alpha})$.

Comments
(1) Comparisons are most interesting under heavy traffic conditions. The variability of P- and NP-message numbers is less extreme, and they are nearer to each other than under a head-of-the-line priority regime.
(2) It is possible to show that as $\alpha \to 1$, $\sigma_X \to r^{\frac{1}{2}}/(1-r)$, the non-priority system value, and as $\alpha \to 0$ it must tend to infinity. There is accordingly at least one value of α for which σ_X is least. For $r = 0.9$, an interesting phenomenon is observed where more than one local minimum occurs.
(3) The values of $\sigma_{W_{np}}$ can be compared with Table 24 of Chapter 4. The variability of W_p and W_{np} is greatly reduced by the alternating priority system though, again, the P-messages are less well favoured than under head-of-the-line priority. Since $\sigma_{W_{np}} \to 1/(1-r)$ in units of mean processing time as $\alpha \to 0$, we again see that a value of α exists for which $\sigma_{W_{np}}$ is minimized, a value which corresponds to assigning priority to a rather large proportion of the total traffic.

References

[1] French, S. (1986) *Decision theory*. Ellis Horwood. Chichester.
[2] Thom, R. (1975) Structural stability and morphogenesis: an outline of a general theory of models, Benjamin. Reading, MA.
[3] Koopman, B. O. (1967) Search and information theory: part of final report on stochastic processes in certain naval operations. New York, NY, Columbia University, Department of Mathematics. [AD 687 543].
[3] Mela, D. F. (1961) Information theory and search theory as special cases of decision theory. *Operations Research*, **9**, 907–909.
[5] Richardson, H. R. (1973) ASW information processing and optimal surveillance in a false target environment. Daniel H. Wagner Associates. Paoli, PA.
[6] Barker, W. H. II. (1977) Information theory and the optimal detection search. *Operations Research*, **25**, 304–314.
[7] Pierce, J. G. (1978) A new look at the relation between information theory and search theory. *In*: Levine, R. D. and Tribus, M. (eds). The Maximum Entropy Formalism. MIT Press. Cambridge, MA.
[8] Conolly, B. W. (1980) Techniques in Operational Research, Vol. 2, Ellis Horwood. Chichester.
[9] Koopman, B. O. (1980) Search and Screening. Pergamon.
[10] Lanchester, F. W. (1916) Aircraft in warfare: The dawn of the fourth arm. Constable. London.
[11] Burg, J. (1970) New concepts in power spectrum estimation. Proc. 40th International SEG Meeting, New Orleans.
[12] Whittaker, E. T. and Watson, G. N. (1958) A course of modern analysis. C.U.P.
[13] Abramowitz, M. and Stegun, I. A. (1964) Handbook of mathematical functions, Dover.
[14] Jaiswal, N. K. (1968) Priority Queues. Academic Press. New York, London.
[15] Conolly, B. W. and Dovletis, G. (1986) Head-of-the-line priority in an exponential queueing system with two service points. University of London, Queen Mary College. Dept. of Computer Science and Statistics. Res. Report. No. 373.

References

[16] Conolly, B. W. and Dovletis, G. (1987) Head-of-the-line priority in an M/M/2 queueing system with two mean service rates. University of London. Queen Mary College, Dept. of Computer Science and Statistics. Res. Report. No. 411.

[17] Pinto, F. de Oliveira and Conolly, B. W. (1981) Applicable mathematics of non-physical phenomena. Ellis Horwood. Chichester.

[18] Morse, P. M. and Kimball, G. E. (1946) Methods of operations research. OEG Report 54, Navy Dept., Washington, D.C.

[19] Birkhoff, G. and MacLane, S. (1965) A survey of modern algebra. Macmillan, New York.

[20] Athans, M. *et al.* (eds.) (1978–1985) Proceedings of MIT/Office of Naval Research Workshops on C3 Systems. Laboratory for Information and Decision Systems. Massachusetts Institute of Technology. Cambridge, Mass. 02139.

[21] Cobham, A. (1954) Priority assignments in waiting line problems. *Operations Research*, **2**, 70–76.

[22] Conolly, B. W. (1975) Lecture Notes on Queueing Systems. Ellis Horwood. Chichester.

[23] Conway, R. W., Maxwell, W. L. and Miller, L. W. (1967) Theory of Scheduling. Addison-Wesley. Reading, MA.

[24] Kleinrock, L. (1976) Queueing Systems. Wiley. New York.

[25] Moscardini, A. O. *et al.* (1988) Mathematical modelling for information technology. Ellis Horwood. Chichester.

Index

ancient warfare, 135, 136
aimed fire, 136, 137, 163,164
area fire, 136, 137, 160, 163, 164

Barker, W. H. II, 9
Bernoulli trial, 18
binary symmetric channel, 57, 59
birth–death process, 10
Burg, J., 10
busy period, 90, 123, 167

catastrophe theory, 7, 8
channels,
  binary, 57, 59
  independent, 77
  parallel, 77
classification, 18
Cobham, A., 165
command and control, 7, 8, 112
communication theory, 56
congestion, 112
control theory, 7, 8
Conway, R. W., 166

decision making, 63, 69
demand, restriction of, 113
detection opportunity, 18
directed fire, 136, 137, 163, 164
discrete Fourier transform, 72

effective coverage of sensor, 57
entropy, 9, 49, 50
  maximum, 9, 10
  source, 59
erasure,
  channel, 56, 57, 59
  probability, 73, 77, 79
Erlans, A. K., 88, 113
exchange ratio, 146, 160, 164

false,
  contact, 9, 55
  target, 13 *et seq.*
Fast Fourier Transform, 72

filter centre, 10, 56
finite Fourier series, 72
'furthest-on' interval, 53
fuzzy,
  logic, 69
  set theory, 7, 8

garbling, 69
generating function, 70, 71, 78, 100, 120
  double, 121
Goodman, R. R., 11
guerrilla force, 160

head-of-the-line priority, 165

information, 50, 52–57, 135–138, 146–150, 160, 162, 164
  average, 52, 53
  decay, 48, 138, 150, 162
  marginal, 52, 53
  mutual, 49, 50, 57, 58
  rate, 54
  self-, 49
  Shannon, 132
  theory, 7–10
  threshold, 160, 162, 164

Jaiswal, N. K., 165

Kleinrock, L., 166
Koopman, B. O., 8

Lanchester, F. W., 10, 135
Lanchester's equations, 135–137, 162, 163
Legendre,
  functions, 72, 77
    asymptotic properties of, 73, 77
  polynomials, 72
Little's formula/result, 97, 131, 166
loss systems, 113

management,
  of demand, 112
  of information flow, 87

by demand adjustment, 88
of priority system, 112, 113
of traffic intensity, 112
Maxwell, W. L., 166
Mela, D. F., 8, 9
merged data, 63
messages,
    transmission of as queueing system, 87
method of balance, 89, 166
Miller, L. W., 166
modern warfare, 136

Pierce, J. G., 9
priority,
    abuse of, 87, 91
    alternating, 118, 165
    head-of-the-line, 87, 165
    preemptive, 165

queueing system, 87 — passim in chapters 5, 6

random walk, 53, 54
reconnaissance, 56, 137
Richardson, H. R., 9
roundabouts, 127

Saclant ASW Research Centre, 11

search,
    effort, 18, 32
    theory, 8
sensor error, 57
self-regulating mechanism, 118
service opportunity, 90
Shannon, C., 8
shift operator E, 95
stalemate, 150, 160, 163, 164
statistical equilibrium, 87
stochastic network, 78
surveillance, 9, 10, 13, 48, 56
symmetric channel, 57
system state, 57, 88
system time, 90, 97, 130

target density,
    posterior, 49
    prior, 48
Thom, R., 8
threshold,
    information, 160, 162, 164
traffic,
    intensity, 99, 112
    lights, 127

vehicle flow, 127

Mathematics and its Applications

Series Editor: G. M. BELL, Professor of Mathematics, King's College London (KQC), University of London

Faux, I.D. & Pratt, M.J.	Computational Geometry for Design and Manufacture
Firby, P.A. & Gardiner, C.F.	Surface Topology
Gardiner, C.F.	Modern Algebra
Gardiner, C.F.	Algebraic Structures: with Applications
Gasson, P.C.	Geometry of Spatial Forms
Goodbody, A.M.	Cartesian Tensors
Goult, R.J.	Applied Linear Algebra
Graham, A.	Kronecker Products and Matrix Calculus: with Applications
Graham, A.	Matrix Theory and Applications for Engineers and Mathematicians
Graham, A.	Nonnegative Matrices and Applicable Topics in Linear Algebra
Griffel, D.H.	Applied Functional Analysis
Griffel, D.H.	Linear Algebra
Guest, P. B.	The Laplace Transform and Applications
Hanyga, A.	Mathematical Theory of Non-linear Elasticity
Harris, D.J.	Mathematics for Business, Management and Economics
Hart, D. & Croft, A.	Modelling with Projectiles
Hoskins, R.F.	Generalised Functions
Hoskins, R.F.	Standard and Non-standard Analysis
Hunter, S.C.	Mechanics of Continuous Media, 2nd (Revised) Edition
Huntley, I. & Johnson, R.M.	Linear and Nonlinear Differential Equations
Jaswon, M.A. & Rose, M.A.	Crystal Symmetry: The Theory of Colour Crystallography
Johnson, R.M.	Theory and Applications of Linear Differential and Difference Equations
Johnson, R.M.	Calculus: Theory and Applications in Technology and the Physical and Life Sciences
Jones, R.H. & Steele, N.C.	Mathematics of Communication
Jordan, D.	Geometric Topology
Kelly, J.C.	Abstract Algebra
Kim, K.H. & Roush, F.W.	Applied Abstract Algebra
Kim, K.H. & Roush, F.W.	Team Theory
Kosinski, W.	Field Singularities and Wave Analysis in Continuum Mechanics
Krishnamurthy, V.	Combinatorics: Theory and Applications
Lindfield, G. & Penny, J.E.T.	Microcomputers in Numerical Analysis
Livesley, K.	Engineering Mathematics
Lord, E.A. & Wilson, C.B.	The Mathematical Description of Shape and Form
Malik, M., Riznichenko, G.Y. & Rubin, A.B.	Biological Electron Transport Processes and their Computer Simulation
Massey, B.S.	Measures in Science and Engineering
Meek, B.L. & Fairthorne, S.	Using Computers
Menell, A. & Bazin, M.	Mathematics for the Biological Sciences
Mikolas, M.	Real Functions and Orthogonal Series
Moore, R.	Computational Functional Analysis
Murphy, J.A., Ridout, D. & McShane, B.	Computation in Numerical Analysis
Nonweiler, T.R.F.	Computational Mathematics: An Introduction to Numerical Approximation
Ogden, R.W.	Non-linear Elastic Deformations
Oldknow, A.	Microcomputers in Geometry
Oldknow, A. & Smith, D.	Learning Mathematics with Micros
O'Neill, M.E. & Chorlton, F.	Ideal and Incompressible Fluid Dynamics
O'Neill, M.E. & Chorlton, F.	Viscous and Compressible Fluid Dynamics
Page, S. G.	Mathematics: A Second Start
Porter, T. & Cordier, J.	Model Formulation Analysis
Prior, D. & Moscardini, A.O.	Model Formulation Analysis
Rankin, R.A.	Modular Forms
Scorer, R.S.	Environmental Aerodynamics
Smith, D.K.	Network Optimisation Practice: A Computational Guide
Shivamoggi, B.K.	Stability of Parallel Gas Flows
Stirling, D.S.G.	Mathematical Analysis
Sweet, M.V.	Algebra, Geometry and Trigonometry in Science, Engineering and Mathematics
Temperley, H.N.V.	Graph Theory and Applications
Thom, R.	Mathematical Models of Morphogenesis
Thurston, E.	Primary Mathematics: Teaching and Learning
Townend, M. S.	Mathematics in Sport
Towend, M.S. & Pountney, D.C.	Computer-aided Engineering Mathematics
Twizell, E.H.	Computational Methods for Partial Differential Equations
Twizell, E.H.	Numerical Methods, with Applications in the Biomedical Sciences
Vince, A. and Morris, C.	Mathematics for Computer Studies
Walton, K., Marshall, J., Gorecki, H. & Korytowski, A.	Control Theory for Time Delay Systems
Warren, M.D.	Flow Modelling in Industrial Processes
Wheeler, R.F.	Rethinking Mathematical Concepts
Willmore, T.J.	Total Curvature in Riemannian Geometry
Willmore, T.J. & Hitchin, N.	Global Riemannian Geometry

Numerical Analysis, Statistics and Operational Research
Editor: B. W. CONOLLY, Professor of Mathematics (Operational Research), Queen Mary College, University of London

Beaumont, G.P.	Introductory Applied Probability
Beaumont, G.P.	Probability and Random Variables
Conolly, B.W.	Techniques in Operational Research: Vol. 1, Queueing Systems
Conolly, B.W.	Techniques in Operational Research: Vol. 2, Models, Search, Randomization
Conolly, B.W.	Lecture Notes in Queueing Systems
Conolly, B.W. & Pierce, J.G.	Information Mechanics: Transformation of Information in Management, Command, Control and Communication
French, S.	Sequencing and Scheduling: Mathematics of the Job Shop
French, S.	Decision Theory: An Introduction to the Mathematics of Rationality
Griffiths, P. & Hill, I.D.	Applied Statistics Algorithms
Hartley, R.	Linear and Non-linear Programming
Jolliffe, F.R.	Survey Design and Analysis
Jones, A.J.	Game Theory
Kapadia, R. & Andersson, G.	Statistics Explained: Basic Concepts and Methods
Moscardini, A.O. & Robson, E.H.	Mathematical Modelling for Information Technology
Moshier, S.	Mathematical Functions for Computers
Oliveira-Pinto, F.	Simulation Concepts in Mathematical Modelling
Ratschek, J. & Rokne, J.	Computer Methods for Global Optimization
Schendel, U.	Introduction to Numerical Methods for Parallel Computers
Schendel, U.	Sparse Matrices
Sehmi, N.S.	Large Order Structural Eigenanalysis Techniques: Algorithms for Finite Element Systems
Späth, H.	Mathematical Software for Linear Regression
Spedicato, E. and Abaffy, J.	ABS Projection Algorithms
Stoodley, K.D.C.	Applied and Computational Statistics: A First Course
Stoodley, K.D.C., Lewis, T. & Stainton, C.L.S.	Applied Statistical Techniques
Thomas, L.C.	Games, Theory and Applications
Whitehead, J.R.	The Design and Analysis of Sequential Clinical Trials

DEC 0 5 1988